Amos Dorinson

Regional Hydrology Fundamentals

Regional Hydrology Fundamentals

RAUL A. DEJU

*Ph.D. New Mexico Institute of Mining
and Technology, Socorro, N.M.*

*Presently Research Geochemist Gulf
Research & Development Company
and Lecturer in Hydrology
University of Pittsburgh
Pittsburgh, Pennsylvania*

GORDON AND BREACH SCIENCE PUBLISHERS
New York   London   Paris

I DEDICATE THIS BOOK
TO MY WIFE LETICIA
AND TO MY SON RAUL

Preface

In recent years interest in the environment and things related to it has grown considerably. This interest stems out from the need to preserve our air, water, and natural resources, and the need to optimize our utilization of these. For many years people took water as a natural thing to have. Today many cities have suffered water shortages, and others have undergone pollution problems, and thus the value of water has increased and taken a new meaning.

This book is concerned with water and attempts to cover in a general fashion the field of hydrology. This is a rather complex and interdisciplinary field covering and encompassing numerous aspects of geology, chemistry, physics, mathematics, sanitation, and civil engineering. This book is designed to aid both the student and the practicing hydrologist to understand more fully the various aspects of this young science.

It is my belief that the most useful approach in solving water management problems is the regional one. A basinwide exploration-evaluation can furnish a more useful picture than isolated area studies. Also, in these modern times one cannot think in terms of supplying water to small areas but must consider the broader problem of most effectively using the waters in the entire basin. This is the task of REGIONAL HYDROLOGY.

Also, one must mention that the problems of water movement, well hydraulics, and regional exploration-evaluation for water are the same problems that petroleum engineers encounter in the study of reservoirs. This book, it is hoped, will serve as a general, simple-minded reference to the student and practicing petroleum engineer trying to understand the problems of oil movement, well flow, and reservoir behavior.

The book is divided into four parts. The first part is concerned with Concepts and Descriptions. Chapter 1 is an Introduction to the field of hydrology and introduces the reader to the concept of HYDROLOGIC CYCLE. Chapter 2 covers the field of surface hydrology and treats the parts of the hydrologic cycle occurring in the earth's surface and the atmosphere. Finally, Chapter 3 is concerned with the concepts of subsurface hydrology and its fundamental laws. Sections 3.10–3.20 are concerned with the mathematical treatment of hydrodynamic laws and can be omitted by those readers who are not mathematically oriented.

Having mastered the concepts and laws of REGIONAL HYDROLOGY the reader can move on to part 2 which discusses the equations involved in describing flow systems. This part can be omitted by those not mathematically inclined who can proceed directly to part 3. Part 2 is divided into two chapters. Chapter 4 gives the mathematical background in potential theory required to solve hydrologic problems. Chapter 5 is concerned with deriving the equations that describe subsurface fluid movement.

Part 3 takes a long, in-depth view at well hydraulics. This is important because a hydrologist must have a clear picture of the movement of underground fluids and also well hydraulics furnishes him the tools required to evaluate underground oil or water reservoirs. Chapter 6 discusses the steady flow of fluid into a well while Chapter 7 is concerned with unsteady flow. The equations derived in Chapter 5 are used in this part in practical hydrologic situations.

The last portion of the book is concerned with the integration of all hydrologic information to solve regional problems. Chapter 8 covers regional subsurface hydrology and includes exploration, drilling and evaluation. Chapter 9 is concerned with regional surface hydrology and discusses the occurrence, distribution, and evaluation of surface water bodies. Chapter 10 is dedicated to the use of geochemical techniques to evaluate water resources in a basin, determine water quality, and analyze pollution.

Three appendices can be found at the end of the book. Since the book uses the metric system, Appendix 1 gives the conversion factors to be used in changing to the English system. Also the empirical equations written in the text using the metric system appear in this appendix in the English system with due reference as to the place where they appear in the text. The metric system has been used since it is the most practical for hydrologic calculations. Appendix 2 gives the conversion factors to change parts per million to equivalents per million. Appendix 3 contains a short description of water laws in the United States.

My main reason for writing this book has been to satisfy the need for a comprehensive treatment of all aspects of REGIONAL HYDROLOGY that would be of use to theoreticians, students, and practicing engineers as well. The book can be used as a general reference or as a text for both petroleum engineering and hydrology courses. The material can be taught at the senior or graduate level for students of engineering and earth sciences. Two semesters are recommended for full coverage of the book.

The book contains a large number of problems for students to solve.

About half of these are theoretical while the other half are practical problems using actual data. Also, the book includes a large number of fully illustrated practical examples.

The mathematical treatment of flow is handled separate from the practical applications and this way a person with less mathematical training can profit from the solutions of flow problems without taking time to derive the equations.

As a suggestion for instructors this book is recommended for the following types of courses:

Introductory Hydrology (1 semester)	Chs. 1–3 and 6.
Theoretical Hydrology (1 semester)	Chs. 1, 3, 6, 7, and 8.
Applied Hydrology (1 semester)	Chs. 1, 3.1–3.8, 6, 7, 8.1–8.2, 9, and 10.
Surface Hydrology (1 semester)	Chs. 1, 2, 3.1–3.8, 9, and 10.
Regional Hydrology (2 semesters)	Chs. 1–10.
Petroleum Mechanics (1 semester)	Chs. 3–8.

Raul A. Deju
Pittsburgh, Pennsylvania
January 5, 1971

Contents

Acknowledgments

The writing of this book has taken several years and many people have rendered their help to the author in putting together information for this book. To all many thanks.

Special thanks should go to Prof. W. C. Ackermann, chief of the Illinois State Water Survey for contributing material for the section on pumping tests. Also many thanks to Dr. William Bertholf who advised me concerning the Water Law Appendix, Mr. John Leighly of the Department of Geography, University of California at Berkeley, Mr. J. F. Friedkin of the International Boundary and Water Commission and to the United States Geological Survey.

The author's Alma Mater, the New Mexico Institute of Mining and Technology, through the courtesy of its President Dr. Stirling A. Colgate, its Bureau of Mines and Mineral Resources Director Mr. Don H. Baker, and the staff in general has contributed textual, tabular, and graphic material for which we gratefully thank them. Thanks are also extended to my colleague W. K. Summers of the Groundwater Dept. at New Mexico Institute of Mining and Technology and to the late Prof. C. E. Jacob also of that department who contributed material used in the writing of this book.

Many thanks are also extended to Dr. Ismael Herrera, Director of the Geophysics Institute of the National University of Mexico in Mexico City for encouraging the author during the many months of writing. Mr. Ignacio Sainz Ortiz, Director of Groundwater of the Ministry of Hydraulic Resources in Mexico is thanked for contributing material and reviewing the section on Evaluation of Groundwater Resources.

Finally, many thanks to Prof. Mahdi S. Hantush, hydrologist, and member of the author's dissertation committee who counseled me as a graduate student and has also always served as my model of an outstanding scientist.

To my wife and son many thanks for bearing with me during the many long nights spent in writing this book.

Raul A. Deju
Pittsburgh, Pennsylvania
January 5, 1971

xvii

NOTE: IN THIS BOOK THE METRIC SYSTEM IS USED THROUGHOUT EXCEPT IN FORMULAS WHERE USAGE OF THE ENGLISH SYSTEM IS SO COMMONPLACE THAT SUCH UNITS MUST BE PRESERVED. FOR THE READER'S CONVENIENCE APPENDIX 1 CONTAINS ALL THE CONVERSION FACTORS NEEDED TO CHANGE FROM ONE SYSTEM TO THE OTHER.

THE AUTHOR

PART 1

CONCEPTS AND DESCRIPTIONS

CHAPTER 1

Introduction

1.1 The Beginnings

Within the field of hydrology one must include four fundamental areas:

(a) groundwater (study of aquifers and subsurface fluids),
(b) potamology (study of rivers),
(c) limnology (study of lakes), and
(d) cryology (study of snow and ice).

In Asia, due to both the aridity of the land and the high population density, the science of water rapidly acquired importance. China developed very early techniques in the art of well drilling. Most of the early wells they drilled were less than 50 meters deep, however, they were well finished. The deepest of the early wells of which there is a record was drilled in China to a depth of 1500 meters nearly 4000 years ago. Unfortunately, the same techniques which were used 4000 years ago are still used in countries like Laos, Cambodia, Thailand, China, and Vietnam.

A very important development was the design and construction of "kanats." These were galleries to collect water percolating through alluvial or sedimentary deposits. Kanats are still used today in parts of Asia and Africa.

It was not until the beginning of the twentieth century that well drilling developed in Europe, and not until this century is there a record of very deep wells drilled in Europe.

1.2 Primitive Theories

The two ancient theories about the origin of water of most importance were:

(a) The Greek theory of Plato. He designed a hydrologic cycle which (see Figure 1.1) included some mysterious canals but still resembles our modern cycle in some aspects.

3

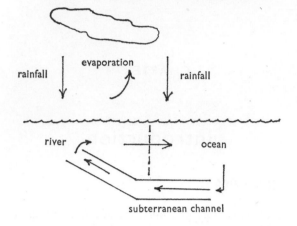

Figure 1.1   The hydrologic cycle according to Plato.

(b) The theory of Marcus Vitruvius who added to Plato's hydrologic cycle the infiltration of water.

The above two and all the other early theories about the hydrologic cycle and the origin of water failed for five fundamental reasons:

(a) The earth does not contain a circuit of underground channels.
(b) The ocean water does not lose all of its salt upon passing through the soil.
(c) The total amount of rainwater is enough to balance the water discharged by rivers and springs.
(d) A large amount of rainwater filters through the soil and replenishes the underground reservoirs.
(e) Capillarity is not a sufficient force to move all the water stored inside the earth.

1.3   Modern Chronology

Numerous scientists have contributed to the advancement of the field of hydrology during the last 150 years. Following a list of some outstanding contributions is given.

Henri Darcy . . . (1856) . . . mathematical law of groundwater flow.

Jules Dupuit ... (1863) ... formula to describe the flow of water to a well.

Adolf Thiem ... (1870) ... modified Dupuit's formula to calculate the characteristics of an aquifer including multiple wells.

Phillip Forchheimer ... (1886) ... introduced the concepts of equipotential surfaces and flow lines.

C. V. Theis ... (1935) ... derived the flow equation to a well for unsteady behavior.

C. E. Jacob ... (1936) ... modified Theis formula.

Morris Muskat ... (1937) ... wrote a book concerning the movement of water through a homogeneous medium.

M. King Hubbert ... (1940) ... derived Darcy's law starting with the flow equations of Navier-Stokes.

Mahdi S. Hantush ... (1948) ... solution of problems involving leaky phases.

P. Ya Polubarinova-Kochina ... (1960) ... theoretical hydrology research.

1.4 The Modern Hydrologic Cycle

The modern hydrologic cycle is far more complicated than the one proposed by our forefathers. The new cycle includes three regions:

(a) The region composed of the atmosphere and the earth's surface where precipitation, evaporation, and runoff take place,

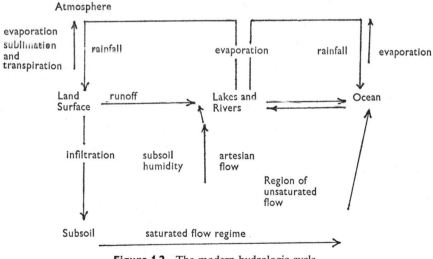

Figure 1.2 The modern hydrologic cycle.

(b) the region of unsaturated groundwater flow, and

(c) the saturated flow region.

Figure 1.2 illustrates the modern hydrologic cycle and includes the three regions mentioned above. This book will be concerned mostly with regions (a) and (c). Region (b) is important to agricultural and soil engineers. However, water in the unsaturated zone is of minimal importance to a hydrologist involved in water-supply work.

CHAPTER 2

Surface Hydrology

The field of surface hydrology is concerned with the study of those parts of the hydrologic cycle that take place either in the atmosphere or on the surface of the earth. Thus, surface hydrology includes the treatment of:

(a) precipitation,
(b) evaporation,
(c) sublimation,
(d) transpiration, and
(e) runoff.

The most fundamental concept in surface hydrology is that of a "basin." The quantitative and/or descriptive study of a basin is an important geomorphologic problem and thus an ideal spot for initiating our discussion.

2.1 Quantitative Geomorphology of a Basin

A river basin can be defined as the total area where runoff in the direction of the river takes place. All points in the basin contribute runoff water to the river and its tributaries. Generally, large basins have a tree shape while small basins usually possess very irregular forms.

The field of geomorphology is concerned with the quantitative study of the basin's physiography. In a quantitative geomorphology study one can work with certain parameters which permit a description of various aspects of a basin. One of these is the drainage density, D, which can be defined by the equation

$$D = \sum L/A \qquad (2.1\text{-}1)$$

where $\sum L$ is the sum of the lengths of all the rivers and tributaries in the basin and A is the drainage area. For example, for the basin of Adobe Creek near Palo Alto, California (Figure 2.1) it was calculated that $\sum L$ is 38.50 km and A is 28.40 km². Thus $D = 1.36$ km⁻¹. Once D is calculated, the type of basin drainage can be calculated using the following chart:

poorly drained basin $\qquad$ $D \leq 0.60 \text{ km}^{-1}$

medium drainage $\qquad$ $0.60 \text{ km}^{-1} < D < 1.50 \text{ km}^{-1}$

excellent drainage $\qquad$ $D \geq 1.50 \text{ km}^{-1}$

Using the above chart the Adobe Creek Basin can be classified as one having medium to good drainage.

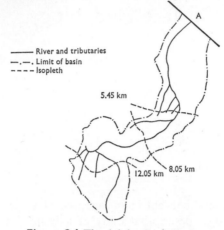

Figure 2.1 The Adobe Creek Basin near Palo Alto, California (taken from R. J. M. DeWiest, 1965).

Other two useful parameters in geomorphologic studies are the order of a river and the order of a basin. A river is said to be of order one if it has no tributaries. A river of order two is one which only has tributaries of order one. Following this reasoning it can be generalized that a river is of order n if all its tributaries are of order $n - 1$ or less. The order of a basin is the same as the order of the main river. The order of a basin depends mainly on four parameters:

(a) the size of the basin,
(b) the amount of drainage,
(c) the geology of the basin, and
(d) the climate.

There is an empirical equation which can be used to obtain the value of the average length of rivers of order n. It states that

$$\bar{l}_n = \bar{l}_1 \bar{l}_{n'} / \bar{l}_{n'-1} \qquad (2.1\text{-}2)$$

where l_1 = average length of rivers of order one, $l_{n'}$ = average length of rivers of order n' (arbitrary), $l_{n'-1}$ = average length of rivers of order $n' - 1$, and l_n = average length of rivers of order n.

If one multiplies l_n by the number of rivers of order n the total length of all rivers of order n is obtained. Doing this for all orders present in the basin and summing the results gives an empirically calculated value for $\sum L$.

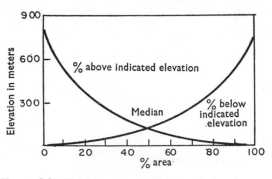

Figure 2.2 Height versus percent area (taken from R. J. M. DeWiest, 1965).

Another type of analysis is the area-elevation study. This type of analysis allows a calculation of the average elevation of the basin. First, one must find the percent of the area of the basin that is over or under a given elevation. Figure 2.2 shows the area-elevation study for the Adobe Creek Basin. From this figure the average basin elevation can easily be obtained.

A slight modification of the area-elevation analysis permits a determination of the age of a basin. First, one must define the relative area A_r and the relative elevation h_r using the formulas

$$A_r = \frac{A_h}{A_{\text{total}}} \quad \text{and} \quad h_r = \frac{h}{h_{\text{max}}}$$

where A_h = area over a given elevation h, A_{total} = total area, h = elevation, and h_{max} = maximum relief contrast. Then a graph of h_r versus A_r is constructed. The result will be one of the three curves in Figure 2.3. Analyses of numerous basins have shown that curve A represents youth, curve B maturity, and curve C old age. The change from A to C represents the intensity of erosion in a basin.

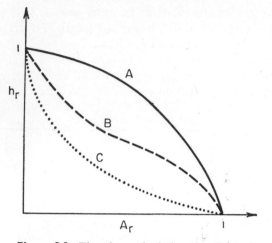

Figure 2.3 Elevation and relative area. Curve A represents a young basin, B a mature one, and C depicts an old basin.

2.2 Descriptive Geomorphology of a Basin

The field of descriptive geomorphology is concerned with the study of all those processes that in one way or another cause changes in the earth's features. Two types of processes need to be considered. These are:

(a) exogen(et)ic processes which are the ones that act from outside the earth's surface such as the climate and erosion, and

(b) endogen(et)ic processes which act from inside the earth's crust, for example, volcanism and orogenesis.

From a hydrologic standpoint the most important of these processes is river erosion. This is an exogenetic process and can be considered as having two phases, which are:

(a) removal of rocks
    (i) rock reduction,
    (ii) material transport,
    (iii) cavitation,
    (iv) abrasion, and

(b) rock accumulation.

The erosive capacity of a river can be evaluated using as index the maximum size of particle of a given type it can carry under the existing flow regime.

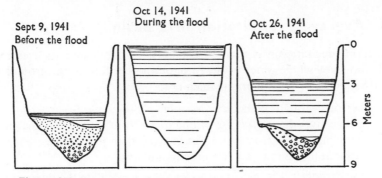

Figure 2.4 Water level changes in the San Juan River caused by the 1941 flood. (Originally from U.S.G.S. Prof. Paper 282B, later shown in Principles of Geology by J. Gilluly and others, W. H. Freeman and Company, copyright 1968).

Another quantity one needs to calculate in order to calculate the erosive capacity of a river is the total particle load it can carry.

The erosive capacity of a river greatly increases during storms and decreases in times of drought. Important changes in the morphology of a basin can occur as the result of intense storms. For example, the case of the San Juan river near Bluff, Utah, shows the drastic changes that took place during the storm of October 1941. Figure 2.4 shows a cross-sectional view of the river prior, during, and after the storm.

A complete analysis of the erosive capacity of a river must also include a study of the process of rock accumulation. The removal and accumulation processes do not occur independently but go side by side. The relation between them is clearly shown for an experimental model of a river in Figure 2.5.

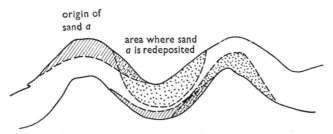

Figure 2.5 Transport and accumulation of sediments in an experimental model of a river. (Originally drawn by J. F. Friedkin. Adapted from Principles of Geology by J. Gilluly and others, W. H. Freeman and Company, copyright 1968.)

2.3   Average Precipitation in a Basin

The term precipitation is used to encompass phenomena such as rainfall, hail, and snow where water in one form or another falls to the ground. There are three types of precipitation which are:

(a) cyclonic- due to planetary and solar phenomena,
(b) convective- due to the rise of humid air from the earth's surface which cools upon reaching a given height and then precipitates, and
(c) orographic- due to air rise near mountains.

To a hydrologist it is important to be able to calculate the average precipitation that has fallen into a given drainage basin over a certain time interval. To calculate average precipitation in a basin one needs first to measure the amount that has fallen in given points of the basin and then one of the following methods can be used.

(A) *Thiessen's method.*   To apply this method, first one must locate the various stations where precipitation has been measured. Then, the adjacent stations can be tied by means of straight lines (see Figure 2.6). Finally, polygons are constructed such that their sides are orthogonal bisectrices of

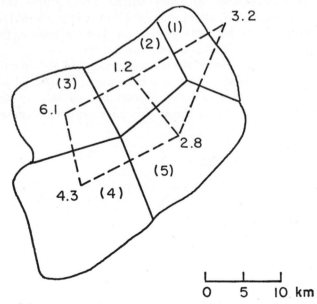

Figure 2.6 Determination of average precipitation using the method of Thiessen.

Table 2.1 Determination of the average precipitation in the basin depicted on Figure 2.6 using the method of the Polygons (Thiessen).

Polygon	Precipitation (cm)	Area (km²)	Prec. × Area
1	3.2	70	224.0
2	1.2	100	120.0
3	6.1	108	658.8
4	4.3	222	954.6
5	2.8	150	420.0
		$\sum 650$	$\sum 2377.4$

$$\text{Average precipitation in the basin} = \frac{\sum P \times A}{\sum A} = 3.6 \text{ cm}$$

the lines connecting adjacent stations. The average precipitation in each polygon is then calculated as indicated in Table 2.1.

(B) *Isohyetal method.* First one traces lines of equal precipitation (isohyets) on a basin map. Then, the average precipitation is calculated for each region between isohyets (see Figure 2.7 and Table 2.2).

(C) *Orographic method.* This method is identical to the previous one except that instead of using isohyets one uses lines of equal elevation above sea level (see Figure 2.8 and Table 2.3).

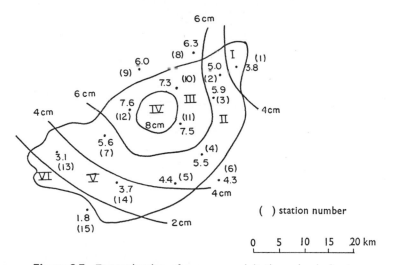

Figure 2.7 Determination of average precipitation using isohyets.

Table 2.2 Determination of the average precipitation in the basin depicted on Figure 2.7 using the isohyetal method.

Region	Area (km²)	Median prec. (cm)	Median prec. × Area
I	31.3	3.8	119
II	320.0	5.1	1630
III	212.0	7.0	1485
IV	50.0	8.0	400
V	191.0	3.4	650
VI	62.2	1.8	112
	$\sum 866.5$		$\sum 4396$

$$\text{Average precipitation in the basin} = \frac{\sum P \times A}{\sum A} = 5.09 \text{ cm}$$

2.4 Precipitation Analysis

Another useful calculation to a hydrologist is a determination of the direction of movement of a storm taking place in a given drainage basin. To do this, one needs to know the cumulative precipitation for various points in the

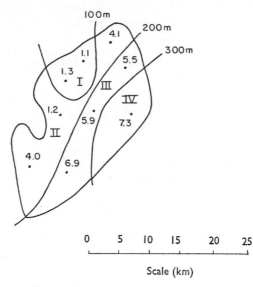

Figure 2.8 Determination of the average precipitation of a basin using the orographic method.

Table 2.3 Determination of the average precipitation in the basin depicted on Figure 2.8 using the orographic method.

Region	Area (km²)	Median prec. (cm)	$P \times A$
I <100 m	75	1.2	90
II 100–200 m	225	4.1	924
III 200–300 m	175	6.1	1007
IV >300 m	125	7.3	912
	$\sum 600$		$\sum 2933$

$$\text{Average precipitation in the basin} = \frac{\sum P \times A}{\sum A} = 4.90 \text{ cm}$$

basin at different times during the storm. For example, for the basin shown in Figure 2.7 we obtained during a 12-hour storm, the data shown on Table 2.4. With these data, the average precipitation in each of the zones limited by isohyets can be calculated (see Figure 2.7). The statistical weight of each station in a zone will depend on the location of the recording station. Table 2.5 shows the results of such an analysis for the basin in question. The hours of most intense storm activity are enclosed in squares. The analysis indicates that it first started to rain in regions III and IV. The storm reached region IV faster than region I and it lasted less in regions I, V, and VI than in regions

Table 2.4 Cumulative precipitation record during a storm in the basin shown on Figure 2.7.

	Stations (S1–S15)														
	Precipitation (cm)														
Time	1	2	3	4	5	6	7	8	9	10	11	12	13	14	15
1 a.m.	0.0	0.0	0.0	0.0	0.0	0.0	0.0	0.0	0.0	0.0	0.0	0.0	0.0	0.0	0.0
2	0.2	0.0	0.0	0.5	0.0	0.0	0.0	0.0	0.0	0.0	0.0	0.0	0.0	0.0	0.0
3	0.6	1.0	1.1	1.5	1.1	0.0	1.0	1.1	1.0	1.0	1.6	1.9	0.0	0.0	0.0
4	1.1	1.6	1.9	1.9	1.9	0.0	1.6	1.6	1.5	1.6	1.9	2.4	2.6	2.5	0.0
5	1.5	1.9	2.3	2.4	2.5	1.0	2.6	2.6	2.7	2.9	2.5	3.5	3.1	3.1	1.8
6	2.5	2.6	2.9	3.4	3.4	2.0	3.5	3.6	3.5	3.1	3.7	4.6	3.1	3.3	1.8
7	3.0	3.3	4.0	4.1	4.1	2.9	3.7	4.0	4.0	4.0	4.6	5.7	3.1	3.7	1.8
8	3.6	4.0	4.9	4.9	4.4	3.9	4.6	4.6	4.6	4.5	5.3	6.8	3.1	3.7	1.8
9	3.8	4.2	5.5	5.5	4.4	4.3	5.6	6.1	6.0	6.1	6.4	7.6	3.1	3.7	1.8
10	3.8	4.9	5.9	5.5	4.4	4.3	5.6	6.3	6.0	6.9	7.1	7.6	3.1	3.7	1.8
11	3.8	5.0	5.9	5.5	4.4	4.3	5.6	6.3	6.0	7.3	7.5	7.6	3.1	3.7	1.8
12	3.8	5.0	5.9	5.5	4.4	4.3	5.6	6.3	6.0	7.3	7.5	7.6	3.1	3.7	1.8

Table 2.5　Determination of the statistically-weighted average precipitation for each zone of the basin on Figure 2.7.

$$I = S1$$
$$II = (S2 + S3 + S4 + S5 + S7)/5$$
$$III = (2\{S10 + S11 + S12\} + S8 + S9)/8$$
$$IV = (S10 + S11 + S12)/3$$
$$V = (S13 + S14)/2$$
$$VI = S15$$

Time	I	II	Region Precipitation (cm) III	IV	V	VI
1 a.m.	0.0	0.0	0.0	0.0	0.0	0.0
2	0.2	0.1	0.0	0.0	0.0	0.0
3	0.6	1.1	1.4	1.5	0.0	0.0
4	1.1	1.8	1.9	2.0	2.5	0.0
5	1.5	2.3	2.9	3.0	3.1	1.8
6	2.5	3.2	3.7	3.9	3.2	1.8
7	3.0	3.8	4.6	4.8	3.4	1.8
8	3.6	4.6	5.3	5.5	3.4	1.8
9	3.8	5.0	6.5	6.7	3.4	1.8
10	3.8	5.3	6.9	7.2	3.4	1.8
11	3.8	5.3	7.1	7.5	3.4	1.8
12	3.8	5.3	7.1	7.5	3.4	1.8

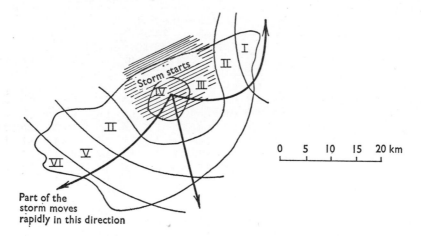

Figure 2.9　Direction in which the storm moves.

II, III, and IV. A possible picture of overall storm behavior is shown in Figure 2.9.

To calculate the average precipitation over a given basin during a storm Horton[1] proposed the following empirical formula

$$P = P_0 \exp \left(-K_1 A^{K_2} \right)$$

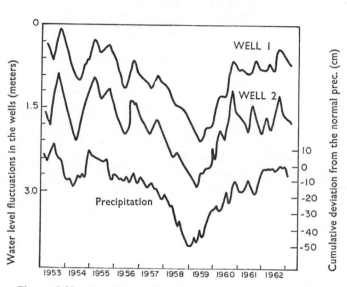

Figure 2.10 Correlation of precipitation and water levels in two wells in Waushara County, Wisconsin, USA. (Taken from W. K. Summers, 1965.)

where P is the average precipitation during a storm over an area A, P_0 is the maximum precipitation at the center of the storm and K_1 and K_2 are constants for a given storm.

Precipitation depends on numerous factors and greatly varies with latitude, elevation, and time of year. It is extremely important to keep good records in selected stations all year around. These data can then be of use in numerous studies. For example, Summers[2] using rainfall data for the period 1953–62 in Waushara County, Wisconsin, found a direct relation between the precipitation pattern and the water levels in wells in the area (see Figure 2.10).

---

[1] Horton, R. E. (1924) *Discussion on Distribution of Intense Rainfall*, ASCE Trans. vol. 87, pp. 578–585.

[2] Summers, W. K. (1965) *Geology and Ground Water Resources of Waushara County, Wisconsin,* U.S.G.S.-WSP 1809-13 U.S. Gov. Printing Office.

2.5 Evaporation, Transpiration, and Sublimation

Evaporation, transpiration and sublimation are three analogous processes that represent water losses to the atmosphere. Evaporation is the loss of liquid water while sublimation is the loss of solid water. Finally, transpiration is the evaporation as executed by plants and trees.

If we consider the earth's globe as a unit we can say that

$$\Delta S + E_0 + E_c = P_0 + P_c \qquad (2.5\text{-}1)$$

where E_0 is the evaporation of ocean water, E_c is the evapotranspiration from the continents, P_0 is the precipitation in the oceans, P_c is the continental precipitation, and ΔS is the change of underground storage.

If instead of considering the earth as a whole only a single continent is analyzed then,

$$P_c = E_c + \Delta S + \text{runoff} \qquad (2.5\text{-}2)$$

Finally if only a basin is treated then

$$\text{Precipitation} + \begin{matrix} \text{Basin} \\ \text{Inflow} \end{matrix} = \text{Evapotranspiration} + \begin{matrix} \text{Change} \\ \text{in} \\ \text{Storage} \end{matrix} + \begin{matrix} \text{Basin} \\ \text{Outflow} \end{matrix}$$

$$(2.5\text{-}3)$$

The three equations above can be called "hydrologic inventories." The first is a global inventory and the other two are respectively a continental and basin inventory. Equations (2.5-1)–(2.5-3) do not include sublimation, however, if a hydrologic inventory is being carried in an area where snowmelt is important, sublimation must be included.

Wüst[3] has estimated that the evaporation from the oceans is roughly 3.28×10^5 km^3 and the continental evapotranspiration is 0.62×10^5 km^3. These values coincide with the earth's overall annual precipitation which is estimated to be 3.90×10^5 km^3. All these figures indicate also the small value of ΔS in relation to global precipitation and evaporation.

There are numerous factors which can affect evaporation and evapotranspiration. The most important are:

(a) temperature,
(b) barometric pressure,
(c) relative humidity,
(d) wind velocity,
(e) atmospheric pollution, and
(f) water pollution.

---

[3] Wüst—see Linsley, R. K., Kohler, M. A., and Paulhus, J. L. (1949) *Applied Hydrology*, McGraw-Hill, New York, 689 pages.

If the temperature rises, the evaporation increases. Pressure produces the opposite effect and increases in pressure diminish the amount of evaporation. Relative humidity increases diminish evaporation. An increase in wind velocity increases evaporation while pollution retards it. Finally, evaporation is easier from oceans, rivers, and lakes than from small surface pools because of the more ready availability of the water.

Meyer[4] has developed an empirical technique to find the evaporation from a given area. He states that:

$$\frac{dE}{dt} = C(e_s - e_a)\left(1 + \frac{W}{10}\right) \tag{2.5-4}$$

where dE/dt = evaporation in inches/day, e_s = saturated vapor pressure, e_a = existing vapor pressure, W = wind velocity in miles/hr at 25 feet over the ground, and C = constant ranging from 0.50 for a humid surface to 0.30 for a 10-meter-deep lake.

Using formula (2.5-4), knowing e_s, e_a, and W and estimating C one can calculate the amount of evaporation over a given time interval.

To experimentally measure evaporation from a lake or dam one uses an evaporometer (see DeWiest[5]). The evaporometer is simply a tank roughly one meter in diameter and 25 cm deep. This tank is filled with water and evaporation is measured. Dr. DeWiest discusses this topic in more detail.

2.6 Runoff

Runoff is that portion of the precipitation that does not filter through the soil or is lost by evaporation but runs through the ground to replenish rivers and lakes. The rivers, however, are not fully dependent on runoff water since they receive a good deal of water through connections with aquifers Also, rivers are sometimes replenished by subsurface runoff. This last phenomenon is of special importance in arid regions where rivers of low discharge drain distant aquifers.

The amount of runoff is highly variable from one region to the next. In the United States (see Figure 2.11) the mean yearly runoff is about 21 cm which is less than one-third the average precipitation in the United States. However, the yearly runoff varies from about 0.50 cm in parts of Arizona and New Mexico to close to 200 cm/year in other parts.

[4] Meyer, A. F. (1942) *Evaporation from Lakes and Reservoirs*, Minnesota Resources Commission, St. Paul, Minnesota.

[5] DeWiest, Roger J. M. (1965) *Geohydrology*, John Wiley & Sons, Inc., N.Y.

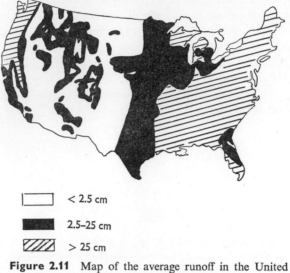

☐ < 2.5 cm

■ 2.5–25 cm

▨ > 25 cm

Figure 2.11 Map of the average runoff in the United
States (data from the U.S.G.S.).

We must distinguish two basic types of runoff:

(a) Immediate runoff, which is the one produced as soon as the water hits
    the ground, and
(b) Delayed runoff which consists of water that first filters through the
    soil and then returns to the ground surface.

There are three types of factors that control the total amount of runoff in
a given area. These are:

(a) Metereological Factors
    (i) rain intensity,
    (ii) temperature,
    (iii) air humidity, and
    (iv) duration of rain.
(b) Topographic Factors
    (i) slope of the ground,
    (ii) extent of the basin,
    (iii) nature of the basin's surface, and
    (iv) physical aspects of the basin's surface.
(c) Geologic Factors
    (i) permeability and specific retention of the soil, and
    (ii) porosity.

The amount of runoff increases as the land slope increases. Compact soils make the movement of water very easy; however, decayed vegetation favors infiltration in lieu of runoff. A permeable soil also favors infiltration and reduces the total runoff. However, if the soil is fully saturated with water, infiltration will not be possible regardless of the permeability of the terrain and then runoff is favored.

Runoff was expressed in Figure 2.11 in terms of cms of water that is runoff volume/area. There are other ways to express runoff, such as, by means of

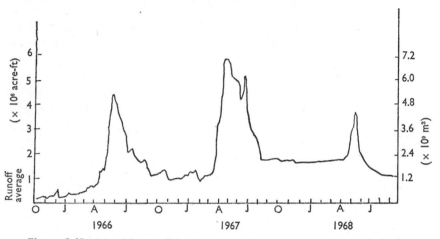

Figure 2.12 Monthly runoff in a hypothetical river during a three year period.

the runoff coefficient which is defined by the equation

$$\alpha = E/\bar{P}$$

where E = amount of runoff in a given point, and $\bar{P}$ = the mean precipitation over the basin.

For example, $\alpha = 0.28$ for the Elba basin and 0.75 for the Rhine. In arid zones α is very low, more so, in cases of intermittent streams. For example, for the various basins in the Sahara desert α lies between 0.02 and 0.07.

The amount of runoff is also expressed sometimes in terms of the runoff volume in acre-ft or in cubic meters. Regardless of the unit in which the measurements are expressed it is necessary that good runoff records be kept as they are useful in the construction of hydrologic balances.

The amount of runoff in a given area varies from month to month. For example, for a hypothetical river basin, the runoff between October 1965 and September 1968 is shown in Figure 2.12. This graph shows that for this basin the maximum runoff is observed in the spring and summer months.

In this case a great deal of the runoff comes from ice melting. Runoff graphs usually run from October to September. This is done in order to have the entire high water period within the same time interval. The United States Geological Survey calls the period October 1–September 30 the *water year*.

2.7 Rivers, Lakes, and Dams

Rivers and lakes have played an important part in the development of human civilization. Western civilization started with an agricultural economy that depended on the rivers Tigris and Euphrates. Even today, rivers and lakes are used as means of transport and as sources of water supply to industries and municipalities.

Water in a river moves by gravity. River velocity is usually small due to the retarding effect of frictional forces. In general, rivers have velocities between 4 and 8 kilometers/hour. Using principles of elementary hydraulics one can readily arrive at the formula

$$V^2 \propto \frac{A}{rp} s \qquad (2.7\text{-}1)$$

where V is the water velocity, A the cross section, s the slope, p the wetted perimeter, and r a roughness coefficient.

The water velocity will also depend on the river cross section. In Figure 2.13 three sections of equal area are shown. Since section I possesses the smallest wetted perimeter it opposses the least friction and thus the water velocity is greater through this section than through the other two.

The total discharge of a river Q is the total amount of water carried by the river past a given point. It can be calculated by the formula

$$Q = \bar{v} A \qquad (2.7\text{-}2)$$

where $\bar{v}$ is the average river velocity and A its cross section. If the velocity is a variable then

$$Q = \int v \, dA \qquad (2.7\text{-}3)$$

If we now consider a symmetrical river (see Figure 2.14) the velocity will be different from point to point in a section. The maximum velocity usually occurs toward the center of the river.

Rivers transport aside from water numerous particles either in suspension or through rolling or saltation. It is important to know the load of suspended particles that a river carries when plans are made to use such river as a source of water supply.

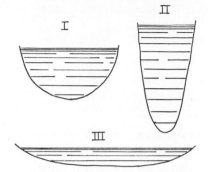

Figure 2.13   Three possible cross sections of a channel. All three have equal areas, however, since section I has the smallest wetted perimeter it opposes the least friction to flowing water. (Originally from U.S.G.S. Prof. Paper 218, later shown in Principles of Geology by J. Gilluly and others, W. H. Freeman and Company, copyright 1968.)

Rivers are a cheap source of water. However, contaminants must be carefully removed before using such water.

Lakes are also important from a water supply standpoint. A lake is simply a large natural depression where water is naturally stored. Due to the lack of movement of lake water it tends to develop planckton and serves as breeding ground for insects and bacteria. Thus, prior to utilization of lake water for any purpose careful sampling operations should be carried out especially in regard to organic materials, bacteria, and planckton.

Dams are also accumulations of surface waters just like lakes except that

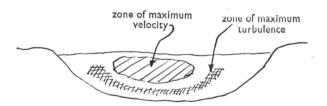

Figure 2.14   Velocity and turbulence distribution in a symmetrical channel. (Originally from an article by J. Leighly in Geographical Review, 1934; adapted from a modified original as appeared in Principles of Geology by J. Gilluly and others, W. H. Freeman and Company, copyright 1968.)

dams are made by man. The construction of dams is a complex engineering problem and requires careful planning. Before a dam is built numerous hydrologic studies should be carried to determine:

(a) the geography of the river to be dammed,
(b) the geologic structure of the area upstream and downstream from the point to be dammed, and
(c) the general stratigraphy of the valley.

Dams, as well as rivers and lakes, are important assets of a region and can greatly enhance the development of a municipality.

There have been numerous studies on the construction of dams and many countries have fostered research in this important field of engineering. A detailed discussion of studies by the Mexican government appears in a publication of the Mexican Ministry of Hydraulic Resources.[6] In Europe some important studies have been carried out by Guizerix and Cornuet[7] concerning the application of radioisotopes to determine escape of water from a dam.

The mathematical theory of the design of a dam is an interesting but complex topic. Some basic fundamentals of the theory of dam design are treated by Muskat.[8]

2.8 Measurement of River Flow by Radioisotopes

The discharge Q from a river is an important parameter in any regional basinwide study. Q can be calculated by conventional means using equations (2.7-2) and (2.7-3) or by radioisotopic techniques such as the dilution, instant injection, and total count methods. Each method will now be discussed in detail.

DILUTION METHOD

This method requires a comparison between the concentration C_1 of a radio-isotope X being introduced into a river at a constant velocity and the concentration C_2 of downstream samples.

Let Q be the discharge, q the rate of injection of isotope X and C_0 its

---

[6] Secretaría de Recursos Hidráulicos de la República Mexicana (1963) *Geohidrología Teórica y Práctica*, publication ZN-1, México, D.F.

[7] Guizerix, J., and Cornuet, R. (1963) *Applications du Sodium-24 à des Mesures de Débits et de Recherches de Fuites*, IAEA, Vienna.

[8] Muskat, M. (1946) *The Flow of Homogeneous Fluids Through Porous Media*, J. W. Edwards, Inc., Michigan, USA.

initial concentration. Then,

$$qC_1 + QC_0 = (Q + q)C_2 \tag{2.8-1}$$

and therefore

$$Q = q(C_1 - C_2)/(C_2 - C_0) \tag{2.8-2}$$

Since, generally $C_2 \ll C_1$ and $C_0 \ll C_2$ equation (2.8-2) reduces to

$$Q = qC_1/C_2 \tag{2.8-3}$$

Thus, Q can be calculated knowing q and measuring C_1 and C_2. A serious drawback of this technique is that it requires sampling of the river over a finite amount of time during which q cannot change.

METHOD OF INSTANT INJECTION

This technique consists in instantly injecting into the river a volume V_1 of radioisotope X of concentration C_1 and measuring C downstream. Consequently,

$$C_1V_1 = \int C \, dV \tag{2.8-4}$$

and since $dV = Q \, dt$ it follows that

$$C_1V_1 = Q \int C \, dt \tag{2.8-5}$$

The activity A of the radioisotope X can be defined as

$$A = C_1V_1 \tag{2.8-6}$$

and combining this equation with (2.8-5) one obtains that

$$Q = \frac{A}{\int C \, dt} \tag{2.8-7}$$

and thus Q can be calculated knowing A and measuring the integral. This technique is cheaper than the dilution method because it does not require continuous injection equipment.

TOTAL COUNT METHOD

This method is a simple modification of the previous one using a geiger counter. The radioactive count N obtained using a geiger counter gives us the concentration of the isotope by means of the equation

$$N = F \int C \, dt \tag{2.8-8}$$

where F is the counter efficiency measured when calibrating the instrument.
Combining equations (2.8-7) and (2.8-8) one obtains that

$$Q = AF/N \tag{2.8-9}$$

and knowing A and F, and measuring N, Q can be readily calculated.

STANDARD METHOD

For continuous measurement of the discharge of a river radioisotopic techniques are not practical and thus it is better in this case to install a permanent discharge gauge station at a point on the river. Then knowing the cross section A and measuring $\bar{v}$, Q can be calculated by the equation

$$Q = \bar{v}A \qquad\qquad (2.8\text{-}10)$$

Numerous such stations exist in important rivers. Mesnier and Iseri[9] discuss the subject in more detail.

2.9   Resource Utilization

Surface waters are in most cases subject to both industrial and human contamination. This pollution of rivers and lakes makes them turbid and their waters tend to acquire colors, odors, and tastes which are not desirable.

Generally surface waters cannot be directly used as municipal water supply sources but must be first treated in order to insure their purity. There are two factors which permit usage of rivers as water supply sources and they are:

(a) low extent of pollution, and
(b) the river is perennial.

Sometimes the water carried by a river contains a certain amount of suspended solids which can be removed using a system of nets and filters. Any particle of sand carried by the water will then be taken out and only pure water will pass the filtering system.

Chemical contamination or pollution can be removed by treatment of the water. The treatment will depend on the type of water and the purity desired. Chlorinization is a common purification process. To obtain very pure water, ion exchange columns can be used, however, they are extremely costly.

Exercises

1. Using Figure 2.1, determine the order of the Adobe Creek basin.
2. Describe various exogenetic processes and discuss their influence in the creation of mountains and valleys.

---

[9] Mesnier, G. N. and Iseri, K. T. (1963) *Selected Techniques in Water Resources Investigations*, USGS-WSP 1669-Z, pp. 1–64.

3. Describe various endogenetic processes and discuss their influence in the shaping of the land surface.

4. Consider a basin (see following figure) with average monthly precipitation as given in the table below. Determine the basin's average precipitation using:

   (a) Thiessen's Method
   (b) Isohyets
   (c) Orographic Method

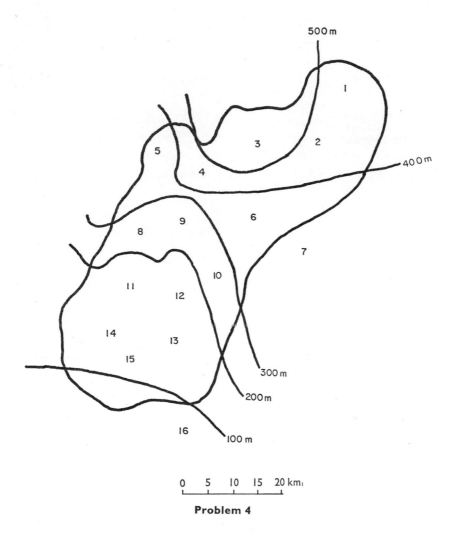

Problem 4

Table

Station	Rainfall (cm)	Station	Rainfall (cm)
1	6.0	9	3.9
2	5.3	10	3.7
3	6.5	11	2.6
4	5.1	12	2.3
5	4.1	13	2.8
6	4.1	14	2.5
7	3.9	15	2.4
8	3.5	16	1.9

5. Consider a hypothetical lake such that $C = 0.40$. Construct a graph of evaporation vs. $(e_s - e_a)$. Extend the graph from $e_s - e_a = 0$ to $e_s - e_a = 0.400$ inches of mercury. Use the following values for W:

(a) $W = 40$ mph
(b) $W = 60$ mph
(c) $W = 100$ mph
(d) $W = 0$.

6. Discuss in detail the possible reasons for the fluctuations of runoff with time for the hypothetical basin depicted in Figure 2.12.

7. What are the most important problems that a hydrologist needs to study when attempting to build a dam in a given area?

8. Discuss and give examples on the best ways of optimizing industrial, agricultural, and municipal uses of water.

CHAPTER 3

Concepts of Subsurface Hydrology

In the previous chapter, the field of surface hydrology was considered in detail. Surface hydrology is the top half of the hydrologic cycle. The bottom half of the cycle is treated in this chapter under the heading of subsurface hydrology.

Subsurface hydrology is concerned with studying hydrologic problems below the land surface. Such problems can be grouped into three basic areas which are:

(i) infiltration,
(ii) unsaturated flow, and
(iii) saturated flow.

Before studying problems of subsurface hydrology, one must study the laws that govern the flow of fluids through a porous medium and be able to apply them to practical situations. This chapter gives the reader an introduction to the fundamental laws of flow and to methods for measuring the basic flow parameters.

3.1 Poiseuille's Law

A basic problem in hydraulics is to describe the flow of a fluid through a tube. This problem was first solved by Poiseuille (1841) and almost simultaneously by Hagen. As we will see in the following section, Poiseuille's law is of great hydrologic importance because the law describing the motion of a fluid through a porous medium (Darcy's law) is a corollary of Poiseuille's law.

First, define the force f needed to impart to a fluid of viscosity μ a velocity gradient dv/dr over an area A parallel to the direction of flow. This equation is

$$f = -\mu A \frac{dv}{dr} \qquad (3.1\text{-}1)$$

29

If the flow region is cylindrical in shape (see Figure 3.1), then

$$f = -\mu 2\pi r l \frac{dv}{dr} \qquad (3.1\text{-}2)$$

and this force is equal but in opposite direction to the pressure $P = F/A$, such that

$$-\mu 2\pi r l \frac{dv}{dr} = P\pi r^2$$

[margin handwritten: confusing notation — of 3.1-1]

and then,

$$dv = -\frac{P}{2\mu l} r\, dr \qquad (3.1\text{-}3)$$

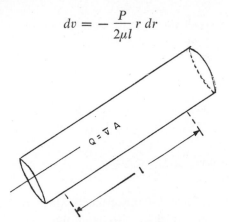

[handwritten on figure: Q = V̄ A]

Figure 3.1   Poiseuille flow.

Integrating (3.1-3) between limits $v = 0$, $r = R$, and $v = v$, $r = r$ we obtain

$$v = \frac{P}{4\mu l}(R^2 - r^2) \qquad (3.1\text{-}4)$$

The total discharge through the cylinder, Q, will then be

[margin handwritten: when h_1 and h_2 are at the top and bottom of the column $Q = \pi R^2 \cdot \frac{\Delta h \rho g R^2}{8\mu l}$]

[left margin handwritten: Q = time rate of volume flow]

$$Q = \int v\, dA = \int_0^R 2\pi r v\, dr = \frac{\pi P R^4}{8\mu l}$$

and since $P = \gamma h$, where γ is the specific mass, it follows that

[right margin handwritten: $Q = \pi R^2 \cdot \frac{\rho g R^2}{8\mu} \cdot \frac{\Delta h}{l}$]

$$Q = (\pi R^2)\frac{\gamma}{8\mu} R^2 \frac{h}{l}$$

The hydraulic gradient is defined as $S^* = h/l$ so that

[right margin handwritten: $\Delta h \rho g$ is the pressure drop in the column of length l]

$$Q = \frac{A\gamma}{8\mu} R^2 S^*$$

Therefore,

$$Q = S^* K' A \qquad (3.1\text{-}5A)$$

[handwritten at bottom: This is a peculiar designation for a quantity whose meaning becomes clear when it is defined as $\gamma = \rho g$ — see p. 32]

where

$$K' = \frac{\gamma R^2}{8\mu}$$ (3.1-5B)

From the above formulas, a macroscopic velocity $\bar{v}$ can be defined and thus

$$\bar{v} = S^* K'$$ (3.1-5C)

Equations (3.1-5A) to (3.1-5C) are known as the law of Poiseuille. This law describes the flow of a fluid through an open pipe. To extend this law to subsurface water flow, one can use the analogy that pores are connected in such a way as to form a set of tubes through which subsurface water flows. Thus, the subsurface flow regime is an interconnected net of microscopic channels and in each of them Poiseuille's law applies.

3.2 Darcy's Law

In 1856, Darcy[1] published his now famous experiment on flow through porous media which led to the development of the law that today carries his name. The apparatus used by Darcy is shown on Figure 3.2. Using this set-up, Darcy was able to conclude that the total discharge Q is given by

$$Q = KA \frac{h_1 - h_2}{dl} = -KA \frac{dh}{dl}$$ (3.2-1)

Confused reasoning and poor notation. In Fig 3.2, l is constant; hence

$$Q = KA\left(\frac{\Delta h}{l}\right)$$

$$\frac{Q}{A} = -\frac{K}{l}\Delta h$$

$$Q/A = \varphi,$$

flux and make h variable. Then

$$\frac{\partial \varphi}{\partial t} = -\frac{K}{l}\frac{\partial h}{\partial t}$$

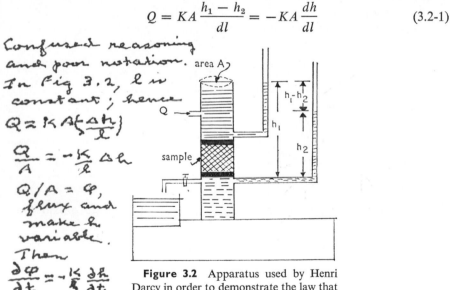

Figure 3.2 Apparatus used by Henri Darcy in order to demonstrate the law that today carries his name.

[1] Darcy, H. (1856) *Les fontaines publiques de la ville de Dijon*, V. Dalmont Paris.

or in terms of the hydraulic gradient

$$Q = -KAS^* \tag{3.2-2}$$

From equation (3.2-2) the macroscopic velocity is found to be

$$\bar{v} = -KS^* \tag{3.2-3}$$

The last two equations found experimentally by Darcy are analogous to equations (3.1-5A) and (3.1-5C) and there lies the correspondence between Darcy's and Poiseuille's laws. The constant K in the above equations is termed hydraulic conductivity or coefficient of permeability and depends on both the nature of the fluid and that of the porous medium.

3.3 The Concepts of Hydraulic Conductivity and Permeability

In the preceding section we saw that Darcy's law involved a constant K called the hydraulic conductivity. Using equation (3.2-1) it is obvious that K has the dimensions of velocity (L/T). The value of K depends on three parameters:

(i) d = average pore diameter
(ii) $\gamma = \rho g$
(iii) μ = viscosity of the fluid.

Thus,

$$K = (\text{constant}) \, (\gamma^{x_1} \mu^{x_2} \, d^{x_3}) \tag{3.3-1}$$

and x_1, x_2, and x_3 can be found by dimensional analysis.

First we must remember that

$$[\gamma] = [ML^{-2}T^{-2}],$$
$$[\mu] = [ML^{-1}T^{-1}],$$
$$[d] = [L], \text{ and}$$
$$[K] = [LT^{-1}].$$

Then, substituting these dimensions into equation (3.3-1) we obtain the dimensional equation

$$[ML^{-2}T^{-2}]^{x_1}[ML^{-1}T^{-1}]^{x_2}[L]^{x_3} = [LT^{-1}]$$

from which

$$x_1 + x_2 = 0$$
$$-2x_1 - x_2 + x_3 = 1, \quad \text{and}$$
$$-2x_1 - x_2 = -1$$

Solving the above system of simultaneous equations, a unique solution is obtained for x_1, x_2, and x_3, and thus

$$K = (\text{constant}) \frac{d^2\gamma}{\mu} \qquad (3.3\text{-}2)$$

Another constant k can be defined which depends only on the nature of the porous medium. This constant is termed intrinsic permeability or simply permeability and is defined by the equation

$$k = Cd^2 = [L^2]$$

Thus,

$$K = \frac{k\gamma}{\mu}$$

and in terms of k, Darcy's law can be given as

$$\bar{v} = -\frac{k\gamma}{\mu} \frac{dh}{dl} \qquad (3.3\text{-}3)$$

Therefore,

$$k = -\mu \frac{Q/A}{dp/dl} \qquad (3.3\text{-}4)$$

The unit of k is the darcy which is defined using equation (3.3-4) as

$$1 \text{ darcy} = \frac{(1 \text{ centipoise})(1 \text{ cm}^3/\text{sec})/(1 \text{ cm}^2)}{(1 \text{ atmosphere/cm})}$$

and roughly 1 darcy $= 10^{-8}$ cm². A reasonably good aquifer will have permeabilities between 10^3 and 10^5 darcies.

The permeability k depends on numerous mineralogical parameters in addition to its dependence on pore diameters. These parameters are usually omitted and are mostly of academic interest. Krumbein and Monk[2] discuss the dependence of the permeability on mineral composition, sphericity, angularity, grain texture, and crystallographic orientation.

3.4 Methods to Measure Hydraulic Conductivity and Permeability

There are several techniques which can be used to measure the hydraulic conductivity and the permeability both in the field and in the laboratory. The apparatus required in such measurements is termed permeameter.

---

[2] Krumbein, W. C. and Monk, G. D. (1943) *Permeability as a function of the size parameters of unconsolidated sand*, AIME Trans. Petroleum Division, vol. 151, pp. 153–163.

When a sample of a sediment is brought to the laboratory so as to measure its permeability the trip affects the sample's porosity, its tightness of packing, and the orientation of the grains. Therefore, laboratory values of hydraulic conductivity and permeability usually differ from corresponding measurements in the field. Also, there is a possibility that the samples taken to the

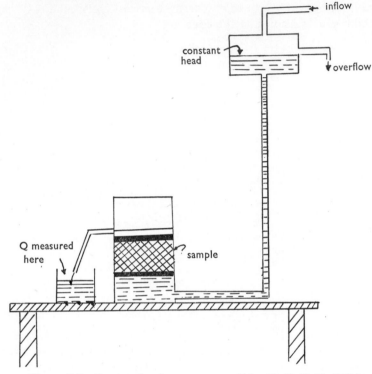

Figure 3.3   Constant head permeameter. (After D. K. Todd, 1959.)

laboratory for permeability determinations may not be fully representative of the sediment being studied.

LABORATORY TECHNIQUES

(A) *Constant head permeameter*

This permeameter can be used to measure the permeability of consolidated or unconsolidated sediments under low heads. The set-up required is based on Darcy's Law and is depicted in Figure 3.3. Water enters the center cylinder through the bottom and is collected in the flask on the left as overflow after having passed through the sample. The hydraulic conductivity is then given by the formula

$$K = - \frac{Q}{A} \frac{dL}{dh} = \frac{VL}{Ath} \qquad Q = V/t$$

where h is the head, A the sample's cross section, L the height of the sample, V the volume of overflow collected in time t and Q the total discharge.

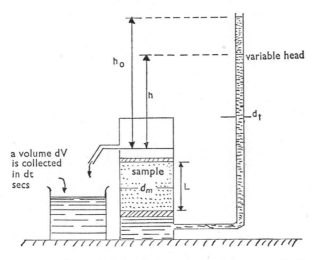

Figure 3.4 Variable head permeameter. (After D. K. Todd, 1959.)

(B) *Lowering Head Permeameter*

In this permeameter water enters from the bottom of the sample proceeding from the tube on the right (see Figure 3.4). As in the previous method water is collected as overflow.

To determine K we have to measure the time required for the water level to fall from one point to another in the tube on the right. Then,

$$K = \frac{d_t^2 L}{d_m^2 t} \ln \frac{h_0}{h}$$

where the symbols are explained in Figure 3.4.

(C) *Sealed Permeameter*

This permeameter has the form of a U-tube and is depicted in Figure 3.5. The sample is located on the U-tube and the water passes through the sample from left to right. Using Darcy's law one obtains that for this permeameter

$$K = \frac{AL}{2at} \ln \frac{h_0}{h}$$

where the symbols are explained in Figure 3.5, and h_0 is the initial head.

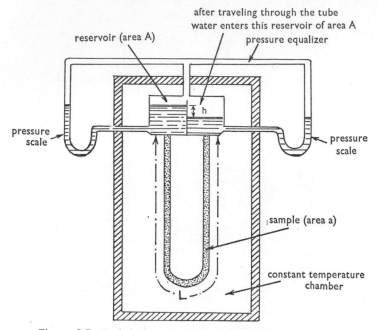

Figure 3.5 Sealed type permeameter. (After D. K. Todd, 1959.)

(D) *Field Methods*

The permeability of a sediment can be measured directly in the field using pumping tests (see Chapters 6 and 7) or by injecting a radioactive solution such as I^{131} and measuring the time required for it to travel from a point "A" to a point "B" in the sediment.

3.5 Porosity

One of the most important properties of a sediment is its porosity. The porosity of a sediment, f, can be defined by the equation

$$f = \frac{\text{pore volume}}{\text{rock volume} + \text{pore volume}} = \frac{V_p}{V_p + V_{\text{rock}}} \qquad (3.5\text{-}1)$$

Some authors like DeWiest[3] also use the void ratio "e" which is defined by the equation

$$e = \frac{V_p}{V_{rock}} \qquad (3.5\text{-}2)$$

Combining (3.5-1) and (3.5-2) one obtains that

$$e = f(1 - f)^{-1} \qquad (3.5\text{-}3)$$

Pores in rocks not only contain water but also contain gases and organic materials. Thus,

$$V_p = V_w + V_g + V_o$$

where V_w = Volume of water, V_g = Volume of gases, and V_o = volume of organic material. Then, the degree of saturation can be defined by the equation

$$S = V_w/V_p \qquad (3.5\text{-}4)$$

Another useful parameter is the volumetric humidity coefficient

$$C = V_w/V_{total} \qquad (3.5\text{-}5)$$

Combining (3.5-1), (3.5-4) and (3.5-5) one obtains that

$$C = Sf \qquad (3.5\text{-}6)$$

From a hydrologic standpoint porosity is a very important thing since movement of subsurface fluids is possible only due to the existence of pores. In a relatively uniform aquifer the permeability can be related to the porosity. Archie[4] has studied the relation between porosity and permeability of sediments specially oil reservoir rocks.

Rocks can be classified in terms of porosity in two groups:

1. Homogeneous porous beds—uniformly distributed pores which make up a large portion of the total framework. Examples: sands and sandy aquifers.
2. Heterogeneous porous beds—few irregularly distributed pores. Examples: carbonates and volcanic rocks.

The amount of water stored in a subsurface aquifer is equal to the total volume of the aquifer multiplied by its porosity. The concept of storage of water or petroleum as will be seen in Section 3.7 is very important.

---

[3] DeWiest, Roger, J. M. (1965) *Geohydrology*, John Wiley & Sons, Inc., N.Y.
[4] Archie, G. E. (1950) *Introduction to petrophysics of reservoir rocks*, American Association of Petroleum Geologists Bulletin, vol. 34, p. 943.

An important calculation is that of porosity of a mixture. For example, consider a homogeneous sand of 40% porosity and gravel of 30% porosity. First, consider the porosity which will result when a cubic meter of sand is mixed with two cubic meters of gravel.

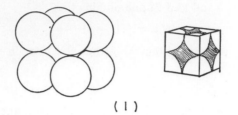

(1)

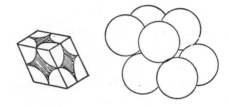

(2)

Figure 3.6 Packing of spheres: (1) cube, and (2) rhombohedron. (From L. C. Graton and H. J. Fraser, 1935, Journal of Geology, volume 43, p. 785, copyright University of Chicago.)

Calculation:

Volume of pores in gravel $= 2 \text{ m}^3 \times 0.30 = 0.60 \text{ m}^3$
Sand left over after filling
all the pores in the gravel $= 1 \text{ m}^3 - 0.60 \text{ m}^3 = 0.40 \text{ m}^3$

$$f = \frac{V_t - V_{rock}}{V_t} = 1 - \frac{V_{rock}}{V_t} = 1 - \frac{(1 \times 0.60) + (2 \times 0.70)}{2 + 0.40}$$

$$f = 0.17$$

Now let us find out what will be the minimum porosity of a mixture of the gravel and the sand being considered. The minimum porosity will occur when one adds sufficient sand to fill the gravel pores. In such case one must add 0.30 m³ of sand for each m³ of gravel. Then,

$$f = 1 - \frac{(0.30 \times 0.60) + (1.00 \times 0.70)}{1.00} = 0.12$$

and therefore the minimum porosity is 12%.

Numerous studies have been carried concerning the packing of spherical grains. Graton and Fraser[5] studied the packing of spheres and concluded

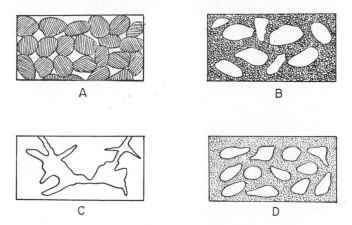

Figure 3.7 Various types of porosities. (A) Well-distributed and highly porous sediment; (B) Gravel mixed with fine sand (low porosity); (C) Limestone that has undergone dissolution; (D) Sedimentary bed with reduced porosity due to mineral deposition at the interstices.

that the maximum porosity occurs for cubical packing (47.64%) while the minimum occurs in the case of rhombohedral packing (25.95%). On Figure 3.6 these two cases are shown. All intermediate cases possess porosities which lie between the cubic and rhombohedral case.

The porosity of a sediment can be affected by geological processes such as compaction, dissolution and cementation. Figure 3.7 shows the effect that these processes have on the porosity.

Other parameters which are intimately related to porosity are the specific yield and the specific retention. The first gives us the part of the total pore volume that yields water when the head is lowered. The specific retention, on the other hand, refers to the volume of trapped pores. Thus,

$$\text{Porosity} = \text{Specific Yield} + \text{Specific Retention} \qquad (3.5\text{-}7)$$

[5] Graton, L. C., and Fraser, H. J. (1935) *Journal of Geology*, vol. 43, p. 785.

EXAMPLE.

In a 250 m² area the head drops 15 m in one year. If the porosity is 0.30 and the specific retention 0.10, calculate the specific yield and the change of storage in cubic meters.

Solution:

$$\text{Volume of the region} = (15 \times 250) \text{ m}^3 = 3750 \text{ m}^3$$
$$\text{where the head decreased.}$$

Using (3.5-7) the specific yield is found to be 0.20. Thus

$$\frac{\text{Storage}}{\text{Change}} = (0.20 \times 3750) \text{ m}^3 = 750 \text{ m}^3$$

There are numerous techniques which can be employed to measure the porosity of a sediment in the laboratory. The classic approach is as follows:

1. Determine the solid's density and weight and calculate the rock volume.
2. Fluid saturate the solid to determine $V_{rock} + V_{fluid}$.

Then,

$$f = V_{fluid}(V_{rock} + V_{fluid})^{-1}$$

3.6 Infiltration to an Aquifer

Part of the precipitation that falls on the land's surface filters through the soil. In geohydrology sometimes one is interested in quantitatively expressing the amount of water filtered into a groundwater reservoir. This will be done in the following section using the concept of storage.

The filtered water can reach two subsurface zones: the aerated and saturated zones (see Figure 3.8). From an agricultural standpoint the aerated zone is important. This zone has a very variable amount of water and movement in it is mostly governed by capillarity.

The saturated zone is the most important from the standpoint of groundwater exploitation. It is separated from the aerated zone by the water table which is a surface maintained at atmospheric pressure. In the saturated zone the movement of water is due to pressure differences and obeys the laws of hydrodynamics.

A porous formation capable of storing and yielding considerable quantities of water is known as an aquifer. Roughly 90% of all aquifers in the world consist of unconsolidated sand and gravel. The most important unconsolidated aquifers are of the quaternary period. For example:

1. alluvial fans composed of sand and gravel,
2. sands and gravels associated with glaciers,
3. coarse sands associated with deltaic deposits, and
4. alluvium.

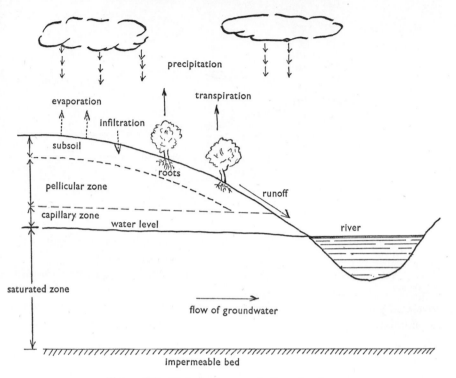

Figure 3.8 Hydrologic zones in the subsurface.

Among the older consolidated aquifers are:

1. permeable sands,
2. fractured lavas,
3. calcareous and dolomitic rocks, and
4. crystalline fractured rocks.

In general aquifers can be classified into three groups:

1. confined or artesian,
2. unconfined or free, and
3. leaky.

Confined or artesian aquifers (see Figure 3.9) are those where ground-water is confined by a pressure greater than atmospheric caused by an impermeable bed overlying the aquifer.

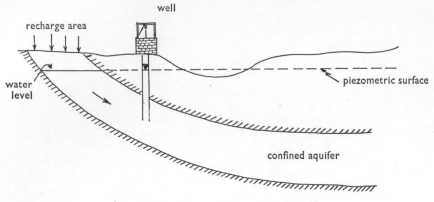

Figure 3.9 Cross-cut of a confined aquifer.

Unconfined aquifers (see Figure 3.10) are those whose upper limit is the water table. These aquifers are also known as free or water table.

The last group of aquifers are the so called "leaky" aquifers. Hantush[6] studied in great detail these aquifers and gave them their name. They are covered by a semipermeable strata which permits a certain amount of water to percolate to them from above, (see Figure 3.11).

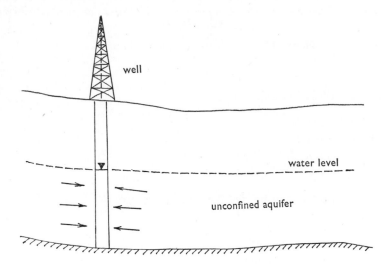

Figure 3.10 Unconfined aquifer.

---

[6] Hantush, M. S. (1949) *Plain Potential Flow of Groundwater with Linear Leakage*, Ph.D. Dissertation, University of Utah.

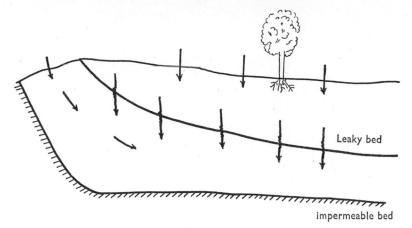

Figure 3.11 Leaky aquifer.

3.7 Transmissivity and Storage

The hydraulic conductivity and the permeability of a material define the capacity of a unit cell of material to transmit water. In order to calculate the amount of water moving through an aquifer as a whole it is necessary to use a coefficient which will be called transmissivity and which can be mathematically defined as

$$T = Kb \qquad (3.7\text{-}1)$$

Basically the transmissivity is the amount of discharge in square meters per day that flows through a vertical unit cut of a confined aquifer of "b" meters thickness under a given hydraulic head and at a 20°C temperature. A simple rule of thumb is that formations with transmissivities of less than 10 m²/day should not be considered as possible aquifers except for small-scale domestic use. If T is greater than 100 m²/day the formation promises to be a reasonably good aquifer.

To completely define an aquifer's physical properties it is also necessary to express quantitatively the amount of water in storage. For this purpose Jacob[7] defined the storage coefficient

First, specific storage can be defined as the amount of stored groundwater that a unit volume of aquifer discharges for each unit of head loss or of piezometric surface change. Thus S_s has dimensions of inverse length $[L^{-1}]$.

---

[7] Jacob, C. E. (1950) *Flow of Groundwater*, in Engineering Hydraulics, ed. Hunter Rouse, John Wiley & Sons, Inc., N.Y.

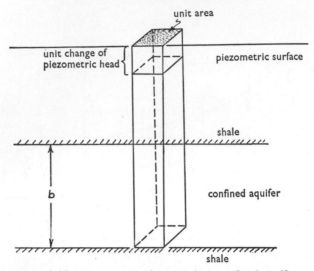

Figure 3.12   The concept of storage in a confined aquifer.

Jacob defined S_s in terms of the fluid and solid compressibilities by the equation:

$$S_s = \rho g(\alpha + f\beta) \qquad (3.7\text{-}2)$$

where ρ is the density of the fluid, g the acceleration of gravity, α the vertical compressibility of the granular skeleton, f the porosity, and β the compressibility of the fluid. Figures 3.12 and 3.13 indicate the meaning of S_s for the case of a confined and an unconfined aquifer.

Having defined S_s, the storage coefficient S of a confined aquifer of thickness "b" can be defined by the equation

$$S = S_s b \qquad \qquad (3.7\text{-}3)$$

where S is nil-dimensional.

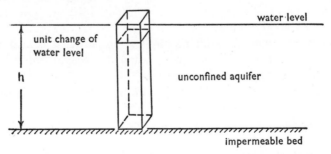

Figure 3.13   The concept of storage in an unconfined aquifer.

For an unconfined aquifer the storage coefficient is equal to the specific yield.

3.8 Types of Flow

In a previous section the three types of aquifers were discussed. The type of flow through these aquifers can be classified from a hydrodynamic standpoint into two groups:

(a) steady flow, and
(b) unsteady flow.

Steady flow is that where $\partial p/\partial t = 0$, or $\partial h/\partial t = 0$ where p is the pressure, h the head, and t the time. In the case of unsteady flow, on the other hand, the pressure and the head vary with time.

Flow can also be referred to as uniform or non-uniform with respect to three dimensional space. Most cases treated in this text will refer to uniform flow.

3.9 Law of Continuity

There are numerous analogies between the laws of hydrodynamics and laws from other branches of physics. For example, the law of continuity is the hydrodynamic principle that expresses mass conservation. It states that in a closed system fluid mass can neither be created nor destroyed.

To obtain this law, consider an element of fluid such as the one depicted in Figure 3.14. For this element,

$$\text{Inflow} = \text{Storage} + \text{Outflow} \qquad (3.9\text{-}1)$$
$$(\Delta I) \qquad (\Delta S) \qquad (\Delta O)$$

The amount of inflow ΔI is

$$\Delta I = \frac{\text{mass}}{\text{time}} = Q\rho = vA\rho$$

and, therefore, for the element of volume in Figure 3.14

$$\Delta I = \rho v_x \, \delta z \, \delta y + \rho v_y \, \delta x \, \delta z + \rho v_z \, \delta x \, \delta y \qquad (3.9\text{-}2)$$

Similarly,

$$\Delta O = \left(\rho v_x + \frac{\partial(\rho v_x)}{\partial x} \, \delta x \right) \delta y \, \delta z + \left(\rho v_y + \frac{\partial(\rho v_y)}{\partial y} \, \delta y \right) \delta x \, \delta z$$
$$+ \left(\rho v_z + \frac{\partial(\rho v_z)}{\partial z} \, \delta z \right) \delta x \, \delta y \qquad (3.9\text{-}3)$$

The amount of stored water is

$$\Delta S = \frac{\partial(\delta M)}{\partial t} = \frac{\partial(\rho f \, \delta x \, \delta y \, \delta z)}{\partial t} = f \frac{\partial \rho}{\partial t} \, \delta x \, \delta y \, \delta z \qquad (3.9\text{-}4)$$

Combining (3.9-1), (3.9-2), (3.9-3), and (3.9-4) we obtain

$$\frac{\partial(\rho v_x)}{\partial x} + \frac{\partial(\rho v_y)}{\partial y} + \frac{\partial(\rho v_z)}{\partial z} = -f \frac{\partial \rho}{\partial t} \qquad (3.9\text{-}5)$$

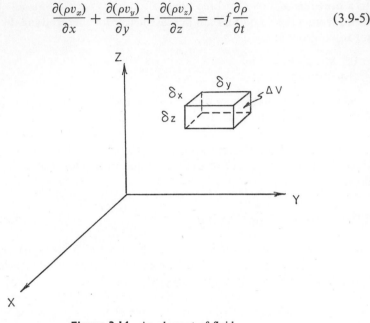

Figure 3.14 An element of fluid.

or in vector form

$$\mathrm{div}\,(\rho \mathbf{v}) = -f \frac{\partial \rho}{\partial t} = \nabla \cdot (\rho \mathbf{v}) \qquad (3.9\text{-}6)$$

This last equation is the law of continuity in vector form.

3.10 Classical Hydrodynamics: The Navier-Stokes Law

In order to solve problems in hydrodynamics, a system of six equations in six variables is used. The law of continuity, discussed in the previous section, contains four variables ρ, v_x, v_y, and v_z. The other two variables required in a hydrodynamic study are the pressure p and the absolute temperature T. Thus, aside from the law of continuity five more equations are needed to complete our hydrodynamic system. Three of these describe the dynamic character of the flow and are known under the common name of "Navier-Stokes" law. The other two equations are the equation of state that describes the nature of the flow and the thermodynamic condition which describes the thermodynamic properties of the fluid. Each of these equations will now be further discussed.

EQUATION OF STATE

This equation is of the form

$$f(p, \rho, T) = 0 \tag{3.10-1}$$

where ρ, p, and T refer to the same element of fluid. For example, the equation of state for an incompressible fluid is $\rho = $ constant.

THERMODYNAMIC CONDITION

This equation is of identical form as the equation of state, that is,

$$g(p, \rho, T) = 0 \tag{3.10-2}$$

Therefore, combining (3.10-1) and (3.10-2) one obtains an equation of the form:

$$h(p, \rho) = 0, \quad \text{or}$$
$$h(p, T) = 0, \quad \text{or}.$$
$$h(\rho, T) = 0.$$

EXAMPLES:

isothermal flow $T = $ constant,

$$\text{adiabatic flow } \frac{T}{T_0} = \left(\frac{\rho}{\rho_0}\right)^{m-1}$$

where m is equal to c_p/c_v and c_p and c_v are the specific heat at constant pressure and constant volume respectively.

NAVIER-STOKES LAW

Consider an element of fluid volume ΔV. This volume is subjected to three forces which are:

(1) Pressure gradients with components $\partial p/\partial x$, $\partial p/\partial y$, and $\partial p/\partial z$.
(2) External forces such as gravity with components per unit volume F_x, F_y, and F_z.

3

(3) Internal forces that oppose fluid movement, such as friction, with components (force/volume):

$$\mu\nabla^2 v_x + \tfrac{1}{3}\mu\frac{\partial\theta}{\partial x}$$

$$\mu\nabla^2 v_y + \tfrac{1}{3}\mu\frac{\partial\theta}{\partial y}$$

$$\mu\nabla^2 v_z + \tfrac{1}{3}\mu\frac{\partial\theta}{\partial z}$$

where μ is the viscosity of the fluid, ∇^2 is the operator

$$\nabla^2 \equiv \frac{\partial^2}{\partial x^2} + \frac{\partial^2}{\partial y^2} + \frac{\partial^2}{\partial z^2}$$

and θ is the dilatation coefficient which is defined by the equation

$$\theta = \text{div } \mathbf{v} = \frac{\partial v_x}{\partial x} + \frac{\partial v_y}{\partial y} + \frac{\partial v_z}{\partial z}$$

Therefore, the sum of the three forces considered gives the total force and applying Newton's law

$$\sum F_s = m\frac{Dv_s}{Dt}$$

one obtains the Navier-Stokes law[8]

$$\rho\frac{Dv_x}{Dt} = -\frac{\partial p}{\partial x} + F_x + \mu\nabla^2 v_x + \tfrac{1}{3}\mu\frac{\partial\theta}{\partial x}$$

$$\rho\frac{Dv_y}{Dt} = -\frac{\partial p}{\partial y} + F_y + \mu\nabla^2 v_y + \tfrac{1}{3}\mu\frac{\partial\theta}{\partial y} \qquad (3.10\text{-}3)$$

$$\rho\frac{Dv_z}{Dt} = -\frac{\partial p}{\partial z} + F_z + \mu\nabla^2 v_z + \tfrac{1}{3}\mu\frac{\partial\theta}{\partial z}$$

The system composed by the law of continuity, the Navier–Stokes law, the equation of state and the thermodynamic condition is mathematically complete, however, it is very difficult to solve.

When one is only interested in a macroscopic treatment of flow the Navier–Stokes law is substituted by Darcy's law which does not involve consideration of individual elements of fluid.

---

[8] Navier, C. L. M. H. (1821) *Ann Chim. phys.*, vol. 19, p. 234.

When the dilatation of the fluid element is negligible, equations (3.10-3) can be written in the form

$$\rho \frac{Dv_i}{Dt} = -\frac{\partial p}{\partial x_i} + \rho g_i + \mu \nabla^2 v_i \tag{3.10-4}$$

where

$i = 1, 2, 3$ and

$i = 1$ corresponds to x,

$i = 2$ corresponds to y, and

$i = 3$ corresponds to z.

Then, substituting ν (the kinematic viscosity) in (3.10-4) one obtains

$$\frac{Dv_i}{Dt} = -\frac{1}{\rho} \frac{\partial p}{\partial x_i} + g_i + \nu \nabla^2 v_i \tag{3.10-5}$$

Now letting V, L, and P represent the characteristic velocity, distance, and pressure respectively and applying the transformations

$$V_i' = v_i/V,$$
$$t' = Vt/L,$$
$$x_i' = x_i/L \quad \text{and} \quad p' = p/P$$

we can reduce equation (3.10-5) to the non-dimensional form

$$\frac{Dv_i'}{Dt'} = \frac{\nu}{VL} \nabla'^2 v_i' + \frac{LG}{V^2}\left(\frac{g_i}{G}\right) - \frac{P}{\rho V^2}\frac{\partial p'}{\partial x_i'} \tag{3.10-6}$$

Since the Reynold's number is $R = VL/\nu$, the Froude number is $F = V^2/LG$ and the cavitation number is $Q^* = P/\rho V^2$

$$\frac{Dv_i'}{Dt'} = \frac{1}{R} \nabla'^2 v_i' + \frac{1}{F}\left(\frac{g_i}{G}\right) - Q^*\left(\frac{\partial p'}{\partial x'}\right)$$

This last equation indicates that when modeling a phenomenon where viscosity plays a more important role than gravity and cavitation the model and the prototype should have the same Reynold's number. Similarly, when gravity or cavitation predominate model and prototype should have the same Froude or cavitation number respectively.

3.11 Bernoulli's Law

The hydrostatic head is a commonly used energy term in hydrodynamics. It can be obtained using Bernoulli's Law. This law states that

$$h = \frac{v^2}{2g} + \frac{p}{\gamma} + z \tag{3.11-1}$$

where h is the hydrostatic head, v the fluid velocity, g the acceleration of gravity, p the pressure, γ the specific mass, and z the elevation.

In a study of groundwater motion, the fluid velocity is very small and, therefore, equation (3.11-1) can be simplified such that

$$h = z + p/\gamma \qquad (3.11\text{-}2)$$

Jacob and other researchers prefer to consider a modified form of equation (3.11-2). They define the hydrostatic head, h, as

$$h = \int_{p_0}^{p} \frac{dp}{\gamma(p)} + z \qquad (3.11\text{-}3)$$

3.12 Darcy's Law in General Form

Darcy's law was expressed in Section 3.2 as

$$v = -K \frac{\partial h}{\partial l}$$

where $l =$ distance in the direction of flow.

Then, in a rectangular coordinate system

$$v_x = -K_x \frac{\partial h}{\partial x}, \qquad v_y = -K_y \frac{\partial h}{\partial y}, \quad \text{and} \quad v_z = -K_z \frac{\partial h}{\partial z} \quad (3.12\text{-}1)$$

where K_x, K_y, and K_z are the hydraulic conductivities in the x, y, and z directions.

In treating petroleum problems, it is more convenient to use pressure rather than head. Then, equations (3.12-1) are best expressed in the form

$$v_x = -\frac{k_x}{\mu} \frac{\partial p}{\partial x}, \qquad v_y = -\frac{k_y}{\mu} \frac{\partial p}{\partial y}, \quad \text{and} \quad v_z = -\frac{k_z}{\mu} \frac{\partial p}{\partial z} \quad (3.12\text{-}2)$$

In order to simplify matters, we can assume to have a homogeneous aquifer with isotropic permeability. Then,

$$v_x = -K \frac{\partial h}{\partial x}, \qquad v_y = -K \frac{\partial h}{\partial y}, \quad \text{and} \quad v_z = -K \frac{\partial h}{\partial z} \quad (3.12\text{-}3)$$

Now let us define a potential function ϕ. Lines of equal ϕ are called equipotentials. As will be seen in the following chapter ϕ satisfies Laplace's equation.

Many potential functions exist for hydrodynamic behavior. Hubbert defined the potential ϕ_H as

$$\phi_H = \int_{P_0}^{P} \frac{dp}{\rho} + gz \qquad (3.12\text{-}4)$$

Another potential function was defined by Bernoulli as

$$\phi_B = \frac{Kv^2}{2g} + K\int_{P_0}^{P} \frac{dp}{\gamma(p)} + Kz \qquad (3.12\text{-}5)$$

In the treatment of problems of groundwater flow, v is very small and, therefore, Jacob was able to simplify (3.12-5) and define a potential ϕ_J as

$$\phi_J = \phi = K\int_{P_0}^{P} \frac{dp}{\gamma(p)} + Kz \qquad (3.12\text{-}6)$$

But, as indicated in the preceding section,

$$h = \int_{P_0}^{P} \frac{dp}{\gamma(p)} + z$$

Then,

$$\phi = Kh, \qquad (3.12\text{-}7)$$

and combining (3.12-3) and (3.12-7) we obtain

$$v_x = -\frac{\partial \phi}{\partial x}, \quad v_y = -\frac{\partial \phi}{\partial y}, \quad \text{and} \quad v_z = -\frac{\partial \phi}{\partial z} \qquad (3.12\text{-}8)$$

As will be shown in the following chapter, the existence of ϕ requires irrotational flow. Even though groundwater flow is (microscopically speaking) rotational, when it is considered macroscopically the rotations statistically balance each other and the motion can be regarded as irrotational. Equation (3.12-8) is the general form of Darcy's law. It can be written in vector form as

$$\mathbf{v} = -\nabla\phi \qquad (3.12\text{-}9)$$

Darcy's law applies only to laminar flow. The Reynold's number $R = dv\rho/\mu$ where d is the grain diameter, v the velocity, and ρ the fluid density defines the region of applicability of Darcy's law. Deviations from this law occur in a porous medium where $R > 1$. The region $R \leq 1$ covers most practically significant cases.

3.13 Equations of Motion

The various equations used to describe common hydrodynamic situations have been discussed in previous sections of this chapter. Now the application of these equations will be discussed using three important situations. In other cases, the equations can be analogously applied.

EXAMPLE 1.

Derive the differential equation that describes the steady isothermal flow of a gas through a porous medium. Repeat for the adiabatic case.

Given:

1. Law of Continuity

$$\nabla \cdot (\rho \mathbf{v}) = -f \frac{\partial \rho}{\partial t} \tag{1}$$

2. Darcy's Law

$$\mathbf{v} = -\nabla \phi \tag{2}$$

3. Equation of State

$$PV = nRT \quad \text{or} \quad \frac{P}{P_0} = \frac{\rho T}{\rho_0 T_0} \tag{3}$$

4. Thermodynamic Condition

$$T = \text{constant} = T_0 \text{ (isothermal)} \tag{4}$$

$$\frac{T}{T_0} = \left(\frac{\rho}{\rho_0}\right)^{m-1} \text{(adiabatic)} \tag{5}$$

5. Gas

$$\phi = Kh = \frac{kp}{\mu} \tag{6}$$

6. Steady State

$$\frac{\partial \rho}{\partial t} = 0 \tag{7}$$

Solution Isothermal Case:

$$\nabla \cdot (\rho \, \nabla \phi) = 0 \qquad \text{(using 1, 2, and 7)}$$

$$\nabla \cdot \left(\rho \frac{k}{\mu} \nabla P\right) = 0 \qquad \text{(using 6)}$$

$$\nabla \cdot \left\{\left[\frac{\rho_0 P T_0}{P_0 T}\right] \frac{k \nabla P}{\mu}\right\} = 0 \qquad \text{(using 3 and 4)}$$

$$\nabla^2 P^2 = 0$$

Solution Adiabatic Case:

$$\nabla \cdot (\rho \nabla \phi) = 0 \qquad \text{(using 1, 2, and 7)}$$

$$\nabla \cdot \left(\frac{\rho k}{\mu} \nabla P\right) = 0 \qquad \text{(using 6)}$$

$$\nabla \cdot \left(\rho_0 \frac{P_0^{1/m}}{P^{1/m}} \nabla P\right) = 0 \qquad \text{(combining 3 and 5)}$$

$$\nabla \cdot (P^{(c_v/c_p)+1}) = 0 \qquad m = \frac{c_p}{c_v}$$

EXAMPLE 2.

Formulate the differential equation which describes the unsteady isothermal flow of an ideal gas moving through an incompressible sand.

Solution: This problem is identical to the first part of the previous one except that

$$\frac{\partial \rho}{\partial t} \neq 0$$

Then

$$-f\frac{\partial \rho}{\partial t} = -f\frac{\partial}{\partial t}\left(\frac{\rho_0 P T_0}{P_0 T}\right)$$

and since $T = T_0$ for isothermal flow

$$-f\frac{\partial \rho}{\partial t} = -\frac{f\rho_0}{P_0}\frac{\partial p}{\partial t}$$

Therefore

$$\tfrac{1}{2}\nabla^2 P^2 = \frac{f\mu}{k}\frac{\partial p}{\partial t}$$

EXAMPLE 3.

Formulate the differential equations that describe the problem of two phase flow of oil and gas through a given incompressible sandy reservoir where a certain fraction of the gas saturates the oil while the remaining gas travels as a separate phase. Assume that the gas and oil velocities are different and that the sand has a different permeability for the oil and gas. Also let the sandy reservoir be fully homogeneous and isotropic and for simplicity's sake assume that the gas is uniformly distributed and expands isothermically when the pressure changes.

Solution: One must start by recalling Darcy's law:

$$v_g = -\frac{k_g}{\mu_g}\nabla P \tag{1A}$$

$$v_l = -\frac{k_l}{\mu_l}\nabla P \tag{1B}$$

where v_l and v_g are the liquid and gas velocities and k_l and k_g their respective permeabilities.

In the case under consideration the equation of continuity must be written (a) for the gas:

$$\nabla \cdot (\rho_g v_g) + \nabla \cdot (Sv_l) = -f\frac{\partial}{\partial t}(S\theta_l + \rho_g\theta_g) \tag{2}$$

where

$\quad\quad\quad S = $ mass of gas dissolved in one unit of liquid,

and

$\quad\quad\quad \theta = $ phase saturation such that $\theta_g + \theta_l = 1 \tag{3}$

(b) for the liquid:

$$\nabla \cdot (\rho_l v_l) = -f \frac{\partial \rho_l}{\partial t}$$

(4)

Combining (1A), (1B) and (2) one obtains

$$\nabla \cdot \left(-\frac{\rho_g k_g}{\mu_g} \nabla P \right) + \nabla \cdot \left(-\frac{S k_l}{\mu_l} \nabla P \right) = -f \frac{\partial}{\partial t} (S\theta_l + \rho_g \theta_g)$$

(5)

Combining (1A), (1B), and (4) one obtains

$$\nabla \cdot \left(-\frac{k_l \rho_l}{\mu_l} \nabla P \right) = -f \frac{\partial \rho_l}{\partial t}$$

(6)

The equation of state for the system under consideration is

$$\frac{P}{P_0} = \frac{\rho_g}{\rho_{0g}} \frac{T_g}{T_{0g}}$$

and the thermodynamic condition can be expressed as

$$T_g = T_{0g}$$

Then, combining the equation of state and the thermodynamic condition one obtains

$$\rho_g = \left(\frac{\rho_{0g}}{P_0} \right) P$$

(7)

By (5) and (7) one gets that

$$\nabla \cdot (C_1 P \nabla P) + \nabla \cdot (C_2 \nabla P) = f \frac{\partial}{\partial t} (C_3 P + \theta_i (S + C_3 P))$$

(8)

where

$$C_1 = \frac{\rho_{0g} k_g}{P_0 \mu_g}, \qquad C_2 = \frac{S k_l}{\mu_l}, \quad \text{and} \quad C_3 = \frac{\rho_{0g}}{P_0}$$

and using (6) one obtains

$$\nabla \cdot \left(\frac{\rho_l k_l}{\mu_l} \nabla P \right) = f \frac{\partial \rho_l}{\partial t}$$

(9)

Since the system discussed has two phases, equations (8) and (9) must be solved simultaneously.

3.14 Differential Form of the Law of Flow Through a Porous Medium

The law of conservation of mass can be expressed as

$$\Delta S = \Delta I - \Delta O$$

(3.14-1)

To begin, consider an elemental volume ΔV in a porous medium (see Figure 3.14). The total inflow entering into ΔV is

$$\rho v_x \, \delta z \, \delta y + \rho v_y \, \delta x \, \delta z + \rho v_z \, \delta x \, \delta y$$

The outflow in the x, y, and z directions is respectively given by

$$\left(\rho v_x + \frac{\partial}{\partial x} (\rho v_x) \, \delta x \right) \delta y \, \delta z,$$

$$\left(\rho v_y + \frac{\partial}{\partial y} (\rho v_y) \, \delta y \right) \delta x \, \delta z, \quad \text{and}$$

$$\left(\rho v_z + \frac{\partial}{\partial z} (\rho v_z) \, \delta z \right) \delta x \, \delta y$$

Also

$$\Delta S = \frac{\partial(\delta M)}{\partial t} \tag{3.14-2}$$

where

$$\delta M = f\rho \, \delta x \, \delta y \, \delta z \tag{3.14-3}$$

and $f =$ porosity.

Since the largest change in element size due to expansion or compression occurs in the z-direction, x and y can be regarded as constants. Then,

$$\frac{\partial(\delta M)}{\partial t} = \left(f \, \delta z \, \frac{\partial \rho}{\partial t} + \rho \, \delta z \, \frac{\partial f}{\partial t} + \rho f \frac{\partial(\delta z)}{\partial t} \right) \delta x \, \delta y \tag{3.14-4}$$

Now let $\alpha =$ vertical compressibility of the sand, $d\sigma_z =$ change in stress, $d(\delta z)/\delta z =$ change in strain, and

$$E = \frac{1}{\alpha} = d\sigma_z \bigg/ \left(\frac{d(\delta z)}{\delta z} \right)$$

Then,

$$d(\delta z) = -\alpha \, \delta z \, d\sigma_z,$$

and differentiating one obtains

$$\frac{\partial(\delta z)}{\partial t} = -\alpha \, \delta z \, \frac{\partial \sigma_z}{\partial t} \tag{3.14-5}$$

The volume of the solid matrix, V_s, can be considered constant. It can be defined by the equation

$$V_s = (1 - f) \, \delta x \, \delta y \, \delta z = \text{constant}$$

Differentiating one obtains

$$dV_s = d\{(1-f)\,\delta x\,\delta y\,\delta z\} = 0$$

$$\delta z\,d(1-f) + (1-f)\,d\,\delta z = 0$$

$$\frac{\partial f}{\partial t} = \frac{1-f}{\delta z}\frac{\partial(\delta z)}{\partial t} \tag{3.14-6}$$

Combining (3.14-5) and (3.14-6) one obtains that

$$\frac{\partial f}{\partial t} = -(1-f)\alpha\frac{\partial\sigma_z}{\partial t} \tag{3.14-7}$$

Before proceeding, let's define the following terms

$$\beta = \text{fluid compressibility,}$$
$$dp = \text{pressure change,}$$
$$d\rho = \text{density change,}$$
$$\rho_0 = \text{original density}$$

Thus,

$$\frac{1}{\beta} = \frac{dp}{(d\rho/\rho_0)}$$

and, therefore,

$$\frac{\partial\rho}{\partial t} = \rho_0\beta\frac{\partial p}{\partial t} \tag{3.14-8}$$

Since the fluid element being considered is in static equilibrium

$$p + \sigma_z = \text{constant}$$

and

$$dp = -d\sigma_z$$

which implies that

$$\frac{\partial p}{\partial t} = -\frac{\partial\sigma_z}{\partial t} \tag{3.14-9}$$

Combining (3.14-7) and (3.14-9) one obtains that

$$\frac{\partial f}{\partial t} = (1-f)\alpha\frac{\partial p}{\partial t} \tag{3.14-10}$$

and combining (3.14-5) and (3.14-9)

$$\frac{\partial\,\delta z}{\partial t} = \alpha\,\delta z\frac{\partial p}{\partial t} \tag{3.14-11}$$

Then, combining (3.14-4), (3.14-8), (3.14-10) and (3.14-11) one gets:

$$\frac{\partial(\delta M)}{\partial t} = (f\rho_0\beta + \rho\alpha)(\delta x\ \delta y\ \delta z)\frac{\partial p}{\partial t} \tag{3.14-12}$$

and combining this last equation with (3.14-1) gives that

$$\left(\frac{\partial\rho v_x}{\partial x} + \frac{\partial\rho v_y}{\partial y} + \frac{\partial\rho v_z}{\partial z}\right)\delta x\ \delta y\ \delta z = (f\rho_0\beta + \rho\alpha)\frac{\partial p}{\partial t}\ \delta x\ \delta y\ \delta z$$

which can be reduced to

$$\rho\frac{\partial v_x}{\partial x} + v_x\frac{\partial\rho}{\partial x} + \rho\frac{\partial v_y}{\partial y} + v_y\frac{\partial\rho}{\partial y} + \rho\frac{\partial v_z}{\partial z} + v_z\frac{\partial p}{\partial z} = (f\rho_0\beta + \rho\alpha)\frac{\partial p}{\partial t}$$

$$\tag{3.14-13}$$

Combining (3.14-13) and Darcy's law leads to

$$\rho K_x\frac{\partial^2 h}{\partial x^2} + K_x\frac{\partial h}{\partial x}\frac{\partial\rho}{\partial x} + \rho K_y\frac{\partial^2 h}{\partial y^2} + K_y\frac{\partial h}{\partial y}\frac{\partial\rho}{\partial y} + \rho K_z\frac{\partial^2 h}{\partial z^2} + K_z\frac{\partial h}{\partial z}\frac{\partial\rho}{\partial z}$$

$$= (f\rho_0\beta + \rho\alpha)\frac{\partial p}{\partial t} \tag{3.14-14}$$

The change in density is, however, very small and thus,

$$\frac{\partial\rho}{\partial x} = \frac{\partial\rho}{\partial y} = \frac{\partial\rho}{\partial z} = 0 \tag{3.14-15}$$

Also since $K_x = K_y = K_z$ for most practical cases (3.14-14) can be reduced to

$$\rho_0 K\nabla^2 h = \rho_0(f\beta + \alpha)\frac{\partial p}{\partial t} \tag{3.14-16}$$

But,

$$p = \gamma h$$

so that (3.14-16) becomes

$$K\nabla^2 h = (f\beta + \alpha)\gamma\frac{\partial h}{\partial t} \tag{3.14-17}$$

Equation (3.14-17) can be simplified using the concept of specific storage (equation 3.7-2). The result is

$$\nabla^2 h = \frac{S_s}{K}\frac{\partial h}{\partial t} \tag{3.14-18}$$

or for confined aquifers

$$\nabla^2 h = \frac{S}{T}\frac{\partial h}{\partial t} \tag{3.14-19}$$

Equation (3.14-18) is the general equation for groundwater flow and is the basis of theoretical hydrology. Jacob[7] is responsible for this derivation which opened the way to the solution of many problems in the theory of groundwater flow. Also this equation establishes the relation between head, storage, permeability, space and time. Jacob continued his work along these lines and extended equation (3.14-18) to cover tidal fluctuations and to include changes in atmospheric pressure. Both of these are discussed in Reference 7 of this chapter. In the next section variations due to atmospheric pressure will be treated and the steps in the mathematical derivation will be outlined.

3.15   The Effect of Changes in Atmospheric Pressure

In the derivation of the preceding section, it was assumed that

$$p + \sigma_z = \text{Total vertical pressure} = \text{Constant}$$

If the atmospheric pressure is regarded as a variable, the above assumption is not valid and one must let

$$\sigma_z = p_a - p + \text{constant}$$

Thus,

$$\frac{\partial \sigma_z}{\partial t} = \frac{\partial p_a}{\partial t} - \frac{\partial p}{\partial t} \qquad (3.15\text{-}1)$$

and instead of using (3.14-9) one applies (3.15-1). Since the resulting equation is extremely difficult to solve, it is recommended that h be found by (3.14-18) and a correction be applied to h so that

$$h_{\text{corrected}} = h + \frac{dh}{dp_a} \qquad (3.15\text{-}2)$$

To find dh/dp_a recall that $dp + d\sigma_z = dp_a$ and since $p = p_a + \gamma h$ it follows that

$$\gamma \frac{dh}{dp_a} = \frac{dp - dp_a}{dp_a}$$

or

$$\gamma \frac{dh}{dp_a} = -\frac{d\sigma_z/dp}{(1 + d\sigma_z/dp)} \qquad (3.15\text{-}3)$$

Now consider a volume V of the aquifer such that $V = V_{\text{water}} + V_{\text{solid}}$. Then

$$dV = dV_{\text{water}} = dV_w$$

and so

$$\frac{dV}{fV} = \frac{dV_w}{V_w} \qquad (3.15\text{-}4)$$

Also,

$$\frac{d\rho}{\rho_0} = -\frac{dV_w}{V_w} \qquad (3.15\text{-}5)$$

where ρ_0 is the original water density.

Combining (3.15-5) with the definition of the compressibility of water one obtains that

$$\frac{1}{\beta} = -\frac{dp}{dV_w/V_w} \qquad (3.15\text{-}6)$$

The compressibility α of the sand can be defined by the equation

$$\frac{1}{\alpha} = -\frac{d\sigma_z}{d(\delta z)/\delta z} = -\frac{d\sigma_z}{dV/V} \qquad (3.15\text{-}7)$$

Then, combining (3.15-4) and (3.15-7) one obtains that

$$-\frac{d\sigma_z}{dV_w/V_w} = \frac{f}{\alpha} \qquad (3.15\text{-}8)$$

and by (3.15-6) and (3.15-8) follows that

$$\alpha \frac{d\sigma_z}{dp} = f\beta \qquad (3.15\text{-}9)$$

Finally, combining (3.15-3) and (3.15-9) leads to

$$\frac{dh}{dp_a} = -\frac{1}{\gamma(1 + \alpha/f\beta)} \qquad (3.15\text{-}10)$$

which is the correction that one needs to apply to the solution of the flow equation when the atmospheric pressure varies.

3.16 General Equation of Flow in an Anisotropic Medium

If the hydraulic conductivities K_x, K_y, and K_z are different then the general flow equation is

$$K_x \frac{\partial^2 h}{\partial x^2} + K_y \frac{\partial^2 h}{\partial y^2} + K_z \frac{\partial^2 h}{\partial z^2} = S_s \frac{\partial h}{\partial t} \qquad (3.16\text{-}1)$$

instead of (3.14-18).

Equation (3.16-1) is harder to solve than (3.14-18), however, by a simple coordinate transformation (3.16-1) can be transformed to the form of (3.14-18).

First, one must define a hydraulic conductivity K such that K be the average of K_x, K_y, and K_z. Then, the following transformation can be applied

$$x = \sqrt{K_x/K}\, x', \qquad y = \sqrt{K_y/K}\, y', \qquad z = \sqrt{K_z/K}\, z'$$

Consequently,

$$\frac{\partial f(x')}{\partial x} = \sqrt{K/K_x}\, \frac{df(x')}{dx'}$$

and

$$\frac{\partial^2 f(x')}{\partial x^2} = \left(\frac{K}{K_x}\right) \frac{d^2 f(x')}{dx'^2}$$

Therefore,

$$\frac{\partial^2 h}{\partial x^2} = \frac{K}{K_x}\frac{\partial^2 h}{\partial x'^2}$$

$$\frac{\partial^2 h}{\partial y^2} = \frac{K}{K_y}\frac{\partial^2 h}{\partial y'^2}$$

and

$$\frac{\partial^2 h}{\partial z^2} = \frac{K}{K_z}\frac{\partial^2 h}{\partial z'^2}$$

Thus, (3.16-1) reduces to

$$\frac{\partial^2 h}{\partial x'^2} + \frac{\partial^2 h}{\partial y'^2} + \frac{\partial^2 h}{\partial z'^2} = \frac{S_s}{K}\frac{\partial h}{\partial t} \tag{3.16-2}$$

Equation (3.16-2) is analogous to (3.14-18) except for the use of coordinates x', y', and z'.

3.17 Flow from a Region of Conductivity K_1 to One with Conductivity K_2

In this section, a fluid moving from one region of hydraulic conductivity K_1 to one of conductivity K_2 will be treated. The mathematical treatment of this problem is analogous to that used when studying the passage of a light ray from a medium with refractive index n_1 to one with index n_2. The only difference is that the law of light refraction involves sines where the refraction of flow lines involves tangents.

The situation being considered is shown on Figure 3.15. The components of the velocity perpendicular to AB are equal in both zones. In other words

$$v_{\perp 1} = v_{\perp 2}$$

and applying Darcy's law one obtains that

$$K_1 \frac{dh_1}{dL_1} \cos \theta_1 = K_2 \frac{dh_2}{dL_2} \cos \theta_2 \qquad (3.17\text{-}1)$$

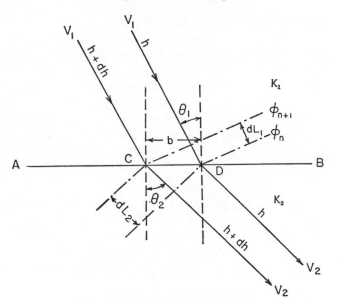

Figure 3.15 Refraction of flow lines.

Also, it is obvious from Figure 3.15 that

$$\sin \theta_1 = \frac{dL_1}{b} \quad \text{and} \quad \sin \theta_2 = \frac{dL_2}{b} \qquad (3.17\text{-}2)$$

Consequently,

$$dL_1 \sin \theta_2 = dL_2 \sin \theta_1 \qquad (3.17\text{-}3)$$

and combining (3.17-1) and (3.17-3) follows that

$$K_1 \frac{dh_1}{\tan \theta_1} = K_2 \frac{dh_2}{\tan \theta_2}$$

and since $dh_1 = dh_2$ we have that

$$\frac{K_1}{K_2} = \frac{\tan \theta_1}{\tan \theta_2}$$

This last equation is the equivalent of Snell's law for the case of flow lines. It gives us a quantitative expression for the change in the direction of motion of groundwater upon encountering a zone of different permeability. For instance, it tells us that if water moves nearly vertically out of a rather impermeable shale, upon reaching a very permeable sand the water will begin to move nearly horizontally.

3.18 Boundary and Initial Conditions

Problems in hydrodynamics involve, as has been seen already, the solution of a partial differential equation under appropriate initial and boundary conditions. For example, the problem of isothermal steady flow of a gas through a porous medium is described by the equation $\nabla^2 P^2 = 0$. The problem of flow of groundwater under unsteady state conditions, on the other hand, requires the solution of the equation

$$\nabla^2 h = \frac{S_s}{K} \frac{\partial h}{\partial t}$$

These equations must be subjected to (n_1, n_2, n_3) boundary conditions with respect to the space coordinates (x, y, z) and m initial conditions. The numbers n_1, n_2, n_3, and m will be the highest orders of the derivatives with respect to x, y, z, and t, respectively, appearing on the differential equation.

Some of the most commonly used boundary and initial conditions are:

(1) the head $h = h_0$ when $t = 0$ h_0 = constant,
(2) the head $h = h_0$ when $t \to \infty$ for all r,
(3) there is an impermeable bed: $K(\partial h/\partial n) = 0$ when $r = r_0$,
(4) $h = h_0$ when $r = r_0$, and
(5) $h = H_w$ when $r = r_w$.

3.19 Analogies with Other Physical Systems

Numerous analogies can be drawn between the parameters used in hydro-dynamics and those used in other fields of physics. These analogies permit us to adapt the equations of heat and electricity flow to the cases of oil and water movement. Also, they allow the use of electrical and/or heat conduction models in the solution of hydrodynamic problems Table 3.1 indicates the basic analogies in oil and water movement, heat flow, electrostatics, and electrodynamics.

Table 3.1 Analogies between physical systems

Oil movement	Water movement	Heat flow	Electrodynamics	Electrostatics
Pressure $= p$	Head $= h$	Temperature $= T$	Voltage $= V$	Potential $= \phi$
Permeability viscosity $\dfrac{k}{\mu}$	Hydraulic conductivity K	Heat conductivity K	Specific conductivity C	Dielectric constant ϵ
The basic law of flow is Darcy's Law	The basic law of flow is Darcy's Law	The basic law of flow is Fourier's Law	The basic law of flow is Ohm's Law	The basic law of flow is Maxwell's Law
$v = -\dfrac{k}{\mu}\nabla P$	$v = -K\nabla h$	$q = -K\nabla T$	$i = -C\nabla V$	$E = -\epsilon\nabla\phi$

3.20 Other Coordinate Systems

All equations discussed in the previous sections were written in cartesian coordinates. Given the symmetry conditions of many problems in hydrodynamics it is easier to solve them in cylindrical or spherical coordinates. For example, the problem of flow to a well located in a radial field is most easily solved by using cylindrical symmetry.

As shown in Section 3.14, the general flow equation is

$$\nabla^2 h = \frac{S_s}{K}\frac{\partial h}{\partial t}$$

If the flow is steady, then $\partial h/\partial t = 0$ and

$$\nabla^2 h = 0 \tag{3.20-1}$$

Since $\phi = Kh$, equation (3.20-1) can be written

$$\nabla^2 \phi = 0 \tag{3.20-2}$$

or in cartesian coordinates

$$\frac{\partial^2 \phi}{\partial x^2} + \frac{\partial^2 \phi}{\partial y^2} + \frac{\partial^2 \phi}{\partial z^2} = 0 \tag{3.20-3}$$

If instead of writing equation (3.20-2) in cartesian coordinates, one desires to use cylindrical coordinates (r, θ, z) the following transformations must be applied:

$$\begin{cases} r = (x^2 + y^2)^{1/2} & \theta = \tan^{-1}\dfrac{y}{x} & z = z \\ x = r\cos\theta & y = r\sin\theta & z = z \end{cases} \tag{3.20-4}$$

where r, θ, and z are related to x, y, and z as indicated in Figure 3.16.

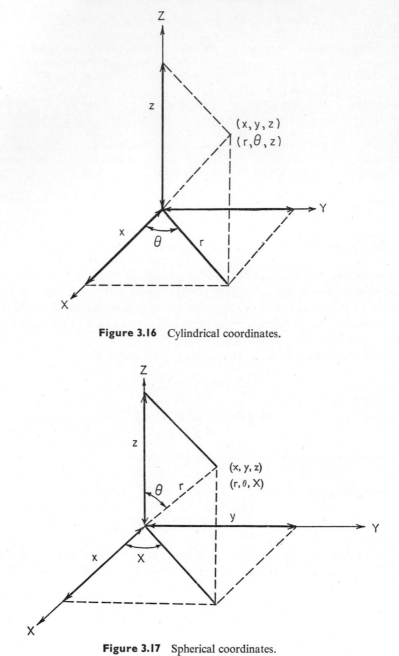

Figure 3.16 Cylindrical coordinates.

Figure 3.17 Spherical coordinates.

Then using (3.20-4), Darcy's law and the steady flow equations can be obtained and so

$$v_r = -\frac{\partial \phi}{\partial r}, \qquad v_\theta = -\frac{1}{r}\frac{\partial \phi}{\partial \theta}, \qquad v_z = -\frac{\partial \phi}{\partial z} \qquad (3.20\text{-}5)$$

and

$$\nabla^2 \phi = \frac{1}{r}\frac{\partial}{\partial r}\left(r\frac{\partial \phi}{\partial r}\right) + \frac{1}{r^2}\frac{\partial^2 \phi}{\partial \theta^2} + \frac{\partial^2 \phi}{\partial z^2} = 0 \qquad (3.20\text{-}6)$$

error +

If spherical coordinates (see Figure 3.17) must be used, then one applies the transformations

$$\begin{cases} r = (x^2 + y^2 + z^2)^{1/2} & \theta = \tan^{-1}\dfrac{(x^2 + y^2)^{1/2}}{z} \\[2mm] \chi = \tan^{-1}\dfrac{y}{x} & x = r \sin\theta \cos\chi \\[2mm] y = r \sin\theta \sin\chi & z = r \cos\theta \end{cases} \qquad (3.20\text{-}7)$$

and the result is

$$v_r = -\frac{\partial \phi}{\partial r}, \qquad v_\theta = -\frac{1}{r}\frac{\partial \phi}{\partial \theta}, \qquad v_\chi = -\frac{1}{r \sin\theta}\frac{\partial \phi}{\partial \chi} \qquad (3.20\text{-}8)$$

and

$$\nabla^2 \phi = \frac{1}{r^2}\frac{\partial}{\partial r}\left(r^2 \frac{\partial \phi}{\partial r}\right) + \frac{1}{r^2 \sin\theta}\frac{\partial}{\partial \theta}\left(\sin\theta \frac{\partial \phi}{\partial \theta}\right) + \frac{1}{r^2 \sin^2\theta}\frac{\partial^2 \phi}{\partial \chi^2} = 0 \quad (3.20\text{-}9)$$

Exercises

1. What range of porosities would have:
   (a) a deposit of mixed sand and gravel?
   (b) a porous limestone formation?
2. Mention a few formations that act as good aquifers.
3. Mathematically speaking what is the difference between steady and unsteady flow?
4. What two additional physical quantities enter into the description of unsteady flow? Illustrate their significance from a geologic point of view.
5. Define briefly:
   (a) free aquifer   (b) specific storage
   (c) transmissivity   (d) porosity
   (e) hydraulic conductivity   (f) volumetric humidity coefficient
   (g) intrinsic permeability   (h) specific retention.

6. What assumptions are made in deriving the equation

$$\nabla^2 h = \frac{S}{T} \frac{\partial h}{\partial t}$$

7. Modify the equation in Problem 6 to steady flow. What additional assumptions must be made?

8. Answer true or false:

    (a) unconsolidated Pleistocene formations generally form good unconfined aquifers
    (b) formations of Cretaceous age never serve as confined aquifers.

9. Derive the differential equation that describes the movement of a gas through a porous medium into a well. The gas behaves as an ideal gas that adiabatically expands. The granular skeleton can be regarded as incompressible.

10. Consider a well field where there are several symmetrically located wells each producing 180 liters of water per second. How long does it take for water to travel the distance between two wells if the porosity is 40% and the wells are 30 meters apart?

11. Suppose one mixes gravel of 36% porosity with sand of 28% porosity. Calculate the resulting porosity when 5 m³ of gravel are mixed with 3 m³ of sand. Determine the void ratio. What proportion of sand and gravel produces minimum porosity and what is this porosity?

12. Consider a coarse gravel and a fine sand both having porosities of 30% and specific yields of 35% and 10% respectively.   Calculate the porosity and the specific yield of a mixture containing

    (a) 10 parts gravel and three parts sand
    (b) 10 parts sand and three parts gravel.

13. The level of water in a certain region is lowered five meters in a year. Laboratory analysis show that the region is a sedimentary zone with a specific yield of 0.20. Calculate the storage change during the year if the total area of the region is 14,000 m²? What is the specific retention of the material if the void ratio is 0.43? What is the storage coefficient?

14. Starting with Darcy's law derive $\nabla^2 h(x, y) = 0$. What flow conditions are represented by this equation?

15. How can one reconcile the validity of Darcy's law with the microscopical rotationality of fluid particles?

16. Compare the law of refraction of flow lines when water passes from a porous medium of hydraulic conductivity K_1 to one of hydraulic conductivity K_2 with Snell's law that describes the passage of light from a region of refractive index n_1 to one of index n_2.

17. Formulate the differential equations that describe the unsteady flow of water through an incompressible sand containing a fraction θ_g of gas. Assume that the gas is uniformly distributed and expands isothermically as the hydraulic head is lowered. The resulting differential equations should be expressed in terms of $h(r, t, \beta, \theta_1, \theta_g)$.

18. Using dimensional analysis determine the form in which the hydraulic conductivity is related to fluid viscosity, grain diameter, and fluid specific weight.

19. In order to apply Darcy's law to a specific case, the flow must be laminar. What index can be used to determine whether or not Darcy's law is applicable to a particular situation?

PART 2

MATHEMATICAL METHODS
AND SYSTEMS

CHAPTER 4

Potential Theory

In the preceding chapter the concepts and laws that serve as a foundation of the field of groundwater hydrology were discussed. These fundamental concepts and laws permit a description of the movement of a fluid through a porous medium. In order to describe such movement, it was necessary to introduce a mathematical function ϕ termed the velocity potential. This function has already been defined in Section 3.12. To completely describe the phenomenon of flow, another function must be defined and related to the velocity potential. This second function is termed the stream function and denoted by ψ. Both, the velocity potential and the stream function come under the field of "Potential Theory." Also these same two functions serve to describe both heat flow and flow of electricity.

This chapter will furnish a brief introduction to potential theory and attempt to show how the velocity potential and the stream function can be applied in groundwater hydrology.

4.1 Movement: Descriptions of Euler and Lagrange

There are two fundamental methods of describing the motion of a fluid through a porous medium. The first is the Lagrangian approach which consists in describing the movement of individual fluid particles. This method is most suitable for microscopic flow analysis and is not convenient for the common macroscopic treatments required in groundwater hydrology.

A second method of describing fluid motion is that of Euler. The Eulerian approach consists in describing the pressure, velocity, and other fluid characteristics in given points or sections of the fluid. This approach is the most useful in groundwater hydrology and is the one used throughout this book.

4.2   Potential Functions

To begin consider a hypothetical fluid moving in a definite direction. A function ψ (the stream function) will now be defined such that its trajectory is tangent at every point in the fluid to the vector representing the fluid's velocity. Thus, in two dimensions

$$\psi = \psi(x, y) \qquad (4.2\text{-}1)$$

and

$$d\psi = \frac{\partial \psi}{\partial x} \, dx + \frac{\partial \psi}{\partial y} \, dy \qquad (4.2\text{-}2)$$

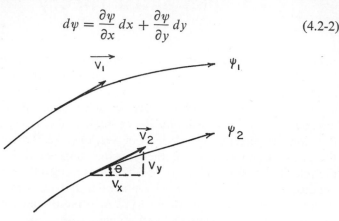

Figure 4.1   Two adjacent streamlines.

If two streamlines (see Figure 4.1) are considered, it follows that

$$\tan \theta = \frac{v_y}{v_x} = \left(\frac{dy}{dx}\right)_{\psi\text{line}} \qquad (4.2\text{-}3)$$

Along any streamline $d\psi = 0$. Therefore,

$$\left(\frac{dy}{dx}\right)_{\psi\text{line}} = -\frac{\partial \psi}{\partial x} \Big/ \frac{\partial \psi}{\partial y}$$

and consequently

$$v_x = -\frac{\partial \psi}{\partial y}$$
$$v_y = \frac{\partial \psi}{\partial x} \qquad (4.2\text{-}4)$$

The velocity potential ϕ can now be introduced. As has been shown in the preceding chapter this function can be defined by the equation

$$\mathbf{v} = -\frac{\partial \phi}{\partial s} \qquad (4.2\text{-}5)$$

where

$$\phi = Kh \qquad (4.2\text{-}6)$$

In cartesian, two-dimensional space

$$\phi = \phi(x, y)$$

and

$$d\phi = \frac{\partial \phi}{\partial x}\, dx + \frac{\partial \phi}{\partial y}\, dy \qquad (4.2\text{-}7)$$

Along an equipotential ($\phi = $ constant) line $d\phi = 0$. Thus,

$$\left(\frac{dy}{dx}\right)_{\phi_{\text{line}}} = -\frac{\partial \phi/\partial x}{\partial \phi/\partial y} = -\frac{v_x}{v_y} \qquad (4.2\text{-}8)$$

Comparing equations (4.2-3) and (4.2-8) it is obvious that streamlines are always perpendicular to equipotential lines.

To complete this discussion of potential functions it is important to determine under which conditions the stream function and the velocity potential exist. To accomplish this consider first the stream function and assume that

$$\frac{\partial^2 \psi}{\partial x\, \partial y} = \frac{\partial^2 \psi}{\partial y\, \partial x}$$

Then,

$$\frac{\partial}{\partial x}\left(\frac{\partial \psi}{\partial y}\right) = \frac{\partial}{\partial y}\left(\frac{\partial \psi}{\partial x}\right)$$

and thus

$$\frac{\partial(-v_x)}{\partial x} = \frac{\partial v_y}{\partial y}$$

which can be written in simpler notation as

$$\operatorname{div} \mathbf{v} = 0 \qquad (4.2\text{-}9)$$

Consequently, the stream function exists only if the flow is divergenceless.

Consider now the velocity potential ϕ and similarly assume that

$$\frac{\partial^2 \phi}{\partial x\, \partial y} = \frac{\partial^2 \phi}{\partial y\, \partial x}$$

Then,

$$\frac{\partial}{\partial x}\left(\frac{\partial \phi}{\partial y}\right) = \frac{\partial}{\partial y}\left(\frac{\partial \phi}{\partial x}\right) \qquad (4.2\text{-}10)$$

and combining (4.2-5) with (4.2-10) one obtains that for the velocity potential to exist it is necessary that

$$\text{curl } \mathbf{v} = 0 \qquad\qquad (4.2\text{-}11)$$

Thus, the velocity potential exists only if the flow is irrotational. Even though groundwater flow is usually microscopically rotational, when one considers it from a macroscopic standpoint by means of Darcy's law, the microscopic rotations balance each other and macroscopically speaking the fluid can be regarded as irrotational. Also the velocity used in Darcy's law must be taken as an average velocity and be defined as

$$\mathbf{v} = \frac{\iiint v \, dx \, dy \, dz}{\iiint dx \, dy \, dz}$$

The velocity potential and the stream function are very useful in describing flow patterns, however, great care must be exercised in applying them only when conditions for their existence are satisfied.

4.3   The Cauchy-Riemann Equations

In the preceding section it was shown that if the functions ϕ and ψ exist they are related to the fluid's velocity by means of the formulas:

$$v_x = -\frac{\partial \phi}{\partial x} = -\frac{\partial \psi}{\partial y} \qquad\qquad (4.3\text{-}1)$$

$$v_y = -\frac{\partial \phi}{\partial y} = \frac{\partial \psi}{\partial x} \qquad\qquad (4.3\text{-}2)$$

and consequently ϕ and ψ satisfy the equations

$$\frac{\partial \phi}{\partial x} = \frac{\partial \psi}{\partial y} \qquad\qquad (4.3\text{-}3)$$

and

$$-\frac{\partial \phi}{\partial y} = \frac{\partial \psi}{\partial x} \qquad\qquad (4.3\text{-}4)$$

Equations (4.3-3) and (4.3-4) are known as the Cauchy-Riemann equations[1] and are of great importance in complex analysis.

---

[1] Ahlfors, Lars V. (1966) *Complex Analysis*, McGraw Hill Book Co., New York, N.Y., 317 pages.

The functions ϕ and ψ can also be regarded as the real and imaginary parts of a function ω which is termed the complex potential and is defined by the equation

$$\omega = \phi + i\psi \qquad (4.3\text{-}5)$$

Some authors prefer to use the complex potential ω instead of the functions ϕ and ψ.

There are two basic mathematical theorems which are of importance in applying the Cauchy-Riemann equations. These are:

Theorem 1. The Cauchy-Riemann equations must be satisfied by the real and imaginary parts of any analytic function.*

Theorem 2. If two harmonic functions (those that satisfy Laplace's equation) ϕ and ψ satisfy the Cauchy-Riemann equations then ψ is said to be the conjugate harmonic function of ϕ. Under the same circumstances ϕ must evidently be the conjugate harmonic function of $-\psi$.

4.4 Application of the Cauchy-Riemann Equations

In the preceding section it was found that under certain conditions the functions ϕ and ψ satisfy the Cauchy-Riemann equations (4.3-3) and (4.3-4). Also Theorem 2 in the preceding section is important because it furnishes a mathematical method to obtain one of the two potential functions knowing the other. To accomplish this, first consider an arbitrary function

$$\theta(x, y) = C \qquad (4.4\text{-}1)$$

The total differential of this function is

$$d\theta = \frac{\partial\theta}{\partial x} dx + \frac{\partial\theta}{\partial y} dy = 0$$

and consequently,

$$\theta = \int \frac{\partial\theta}{\partial x} dx + \int \frac{\partial\theta}{\partial y} dy = C$$

Along an equipotential line ϕ satisfies condition (4.4-1) and along a streamline ψ satisfies (4.4-1) also. Thus, under these conditions

$$\phi(x, y) = C_1$$

and

$$\psi(x, y) = C_2$$

---

* An analytic function in a region R can be defined as a complex function of a complex variable $f(z)$ which is defined and possesses a derivative at each and every point in R.

Then,

$$d\phi = \frac{\partial \phi}{\partial x} dx + \frac{\partial \phi}{\partial y} dy,$$

and

$$d\psi = \frac{\partial \psi}{\partial x} dx + \frac{\partial \psi}{\partial y} dy$$

From these last two equations ϕ and ψ can be found by integration, and therefore,

$$\phi = \int \frac{\partial \phi}{\partial x} dx + \int \frac{\partial \phi}{\partial y} dy, \qquad (4.4\text{-}2)$$

and

$$\psi = \int \frac{\partial \psi}{\partial x} dx + \int \frac{\partial \psi}{\partial y} dy \qquad (4.4\text{-}3)$$

Combining (4.4-2), (4.3-3), and (4.3-4) one obtains that

$$\phi = \int \frac{\partial \psi}{\partial y} dx - \int \frac{\partial \psi}{\partial x} dy \qquad (4.4\text{-}4)$$

Equation (4.4-4) permits a determination of the velocity potential ϕ if the stream function ψ is known.

Similarly, combining (4.4-3), (4.3-3), and (4.3-4) gives that

$$\psi = -\int \frac{\partial \phi}{\partial y} dx + \int \frac{\partial \phi}{\partial x} dy \qquad (4.4\text{-}5)$$

Equation (4.4-5) permits a determination of the stream function if the velocity potential is known.

Three simple examples that illustrate the method discussed above will now be given.

EXAMPLE 1.

Given that $\phi = x^2 y = C_1$ find ψ.

Solution: Using (4.4-5) one obtains that $\psi = -\int x^2 dx + \int 2xy \, dy = C_2$ and therefore

$$\psi = -\frac{x^3}{3} + xy^2 = C_2$$

EXAMPLE 2.

Show that if the velocity potential ϕ for the case of unidirectional steady flow is given by the equation $\phi = v_0 x$ then $\psi = v_0 y$.

Solution: It is given that $\phi = v_0 x$ where v_0 is taken as constant. Therefore $\partial \phi / \partial x = v_0$ and $\partial \phi / \partial y = 0$ and applying equation (4.4-5) one readily obtains that $\psi = \int v_0 \, dy = v_0 y$.

EXAMPLE 3.

Given that $\psi = 2xy = C_1$ find ϕ and determine the value of the velocity vector $\mathbf{v}$ at the points $(1, 0)$ and $(1, 1)$.

Solution: Using equation (4.4-4) one obtains that:

$$\phi = \int 2x \, dx - \int 2y \, dy = C_2$$

$$\phi = x^2 - y^2 = C_2$$

In order to find $\mathbf{v} = v_x\hat{i} + v_y\hat{j}$ one simply applies (4.3-1) and (4.3-2). Thus

$$\mathbf{v} = -2x\hat{i} + 2y\hat{j}$$

and at the point $(1, 0)$:

$$\mathbf{v}(1, 0) = -2\hat{i}$$

and at the point $(1, 1)$:

$$\mathbf{v}(1, 1) = -2\hat{i} + 2\hat{j}$$

4.5 Flow Nets

It has been shown in the preceding sections that the stream function and the velocity potential form an orthogonal system, that is, equipotential lines are perpendicular to streamlines. The graphical representation of the equipotential lines and streamlines of a flow regime forms a flow net. Flow nets are very useful in that they furnish at a single glance the fluid's velocity at every point in the net and they also help in the solution of complex flow problems.

To begin, consider a simple case (see Figure 4.2). This diagram illustrates a flow net. The area between two adjacent streamlines constitutes a flow tube and each rectangle is a flow cell.

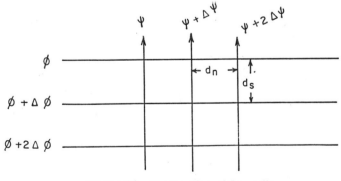

Figure 4.2 Construction of flow cells.

Consider a thickness of unity, then,

$$Q = \mathbf{v} \, dn \qquad (4.5\text{-}1)$$

but since $\mathbf{v} = -\partial\phi/\partial s$ it follows that

$$Q = -\frac{\partial\phi}{\partial s} \, dn. \qquad (4.5\text{-}2)$$

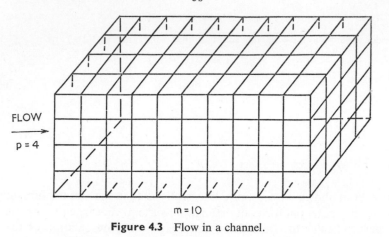

FLOW

p = 4

m = 10

Figure 4.3   Flow in a channel.

Thus, one can say that in a flow tube

$$dq = -K\frac{dh}{ds} \, dn \qquad (4.5\text{-}3)$$

If the flow net is constructed so that all cells are squares then

$$dq = K \, dh \qquad (4.5\text{-}4)$$

Now, if the flow net is considered to be composed of m squares cutting two adjacent streamlines and if the net contains p flow tubes, it follows that

$$Q = p \, dq = K\frac{ph}{m} \qquad (4.5\text{-}5)$$

The last formula is very useful in calculating the discharge Q from channels and pipes. Consider, for example, a rectangular channel such as is shown in Figure 4.3. In this case $p = 4$ and $m = 10$. Thus,

$$Q = 0.40Kh$$

If barriers are placed in the same channel (see Figure 4.4) then the flow net is slightly distorted and $p = 4$ and $m = 13$. Consequently,

$$Q = 0.31Kh$$

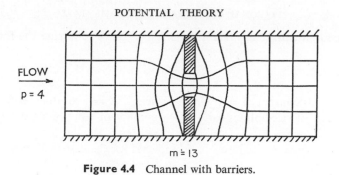

Figure 4.4 Channel with barriers.

Flow nets also allow to qualitatively visualize a given flow regime. For example, consider a 90° corner such as is shown in Figure 4.5. The line OB divides the angle COD into two equal angles COB and BOD and acts as a mirror. The equipotential lines are asymptotic to the walls CO and OD. The velocity of the fluid is infinite at any point where two or more equipotential lines converge (singularities). In general the velocity of the fluid is inversely proportional to the separation between adjacent equipotential lines.

4.6 Calculation of the Amount of Flow Through a Section of an Aquifer Using Flow Nets

Flow nets can be used to calculate the discharge Q through a section of a given aquifer. For example, consider a section of length L of a confined, horizontal, uniformly-thick aquifer (see Figure 4.6). To determine Q in this case one begins by drawing in a scaled sketch of the aquifer the streamlines

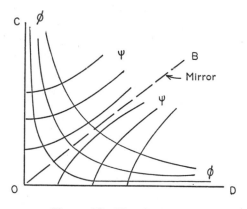

Figure 4.5 Flow in a corner.

and the equipotential lines, such that, $\Delta\phi = \Delta\psi$. Then using the formula

$$Q = Kh\frac{n_\psi}{n_\phi}$$

where Q = discharge, h = hydraulic head, K = hydraulic conductivity, n_ψ = number of streamlines, and n_ϕ = number of equipotentials, Q can be calculated.

Using the above procedure it is found that for the case in Figure 4.6, $Q = 0.40Kh$.

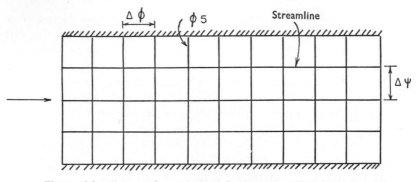

Figure 4.6 Flow net for a confined, horizontal, uniformly thick aquifer.

The same analysis can be applied to aquifers of more complex shape (see Exercise 10 at the end of the chapter).

Exercises

1. Under which conditions of fluid movement does the Stream Function exist? Show mathematically how this function is related to the velocity components.
2. Under which conditions does the Stream Function satisfy Laplace's equation in two dimensions? Repeat the problem in three dimensions.
3. When does the function ϕ exist? Show the relation between this function and the velocity components.
4. Given that $\phi = \phi(x, y)$ and that $v_x = -\partial\phi/\partial x$, and $v_y = -\partial\phi/\partial y$ what condition is necessary for ϕ to satisfy Laplace's equation?
5. What is the basic difference between the Lagrangian and the Eulerian descriptions of fluid movement?
6. Mathematically speaking, what are the implications of assuming that

$$\frac{\partial^2\phi}{\partial x\,\partial y} = \frac{\partial^2\phi}{\partial y\,\partial x}\,?$$

7. Show conclusively that the Cauchy-Riemann equations must be satisfied by the real and imaginary parts of any analytic function.

8. Given that $\phi = x^3 y^2 = C_1$ find the Stream Function and calculate the magnitude of the velocity vector at the point $(1, 0)$.

9. If $\psi = v_0 y^2$ find the corresponding value of ϕ.

10. Consider the three scaled drawings of aquifers shown below and find the discharge Q if the hydraulic conductivity of the aquifers is 5×10^{-2} cm/sec and the hydraulic head in the three cases is 10 mts.

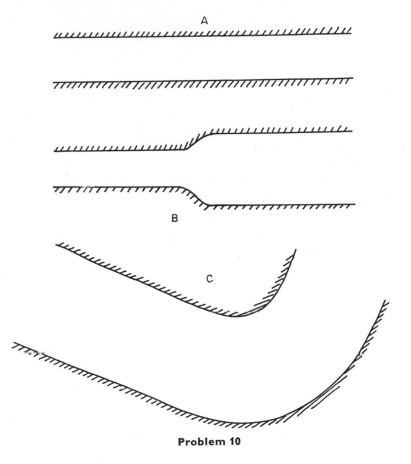

Problem 10

11. What is a singularity? Give examples of points in a flow regime where the velocity is (a) zero, and (b) infinite.

12. Discuss in detail various applications of flow nets not discussed in this chapter.

4

CHAPTER 5

The Flow Equations

In the preceding chapters it was shown that in order to adequately describe the flow of groundwater it is convenient to use the potential function ϕ or the hydraulic head h. These are related by the equation

$$\phi = \frac{p}{\gamma} + h$$

The flow equations expressed in terms of these functions depend on the type of aquifer being considered and on the properties of the beds confining the aquifer.

In this chapter the equations that describe the movement of groundwater in various typical and important cases will be derived and discussed. Emphasis will be given here to the factors to be considered and assumptions to be made in deriving these equations. In later chapters these differential equations will be used to solve various practical problems.

The methods used in setting up the various equations should not be regarded as purely mathematical exercises but due regard should be given to the physical conditions of the particular problem.

Jacob[1] in his study of groundwater flow treats the problem of flow in sands of uniform and non-uniform thickness and applies the results to practical cases. DeWiest[2] has considered the modification that one must make to the Laplace equation to account for leakage. Hantush[3] considers leaky artesian systems, sloping sands and numerous other cases. All these equations form the basis of theoretical hydrology and should be regarded as necessary general background for anyone interested in studying or doing research

---

[1] Jacob, C. E. (1950) *Flow of Groundwater* in Engineering Hydraulics, ed. Hunter Rouse, John Wiley & Sons, Inc., N.Y.

[2] DeWiest, Roger J. M. (1965) *Geohydrology*, John Wiley & Sons, Inc., N.Y.

[3] Hantush, M. S. (1964) *Hydraulics of Wells* in Advances in Hydroscience, ed. V. T. Chow, Academic Press, New York.

in theoretical hydrology. All the above cases and treatment of inclined, non-uniform, and water-table aquifers will be included in this chapter to give the reader an introduction to the mathematical methods necessary in developing a mathematical representation of a hydrologic problem.

Those readers interested only in applied hydrology can skip this chapter and proceed directly to Chapter 6.

5.1 Equation of Flow for a Confined Horizontal Aquifer of Uniform Thickness

We will begin by considering the simplest case, that is, a uniformly thick, confined aquifer of thickness b (see Figure 5.1). In such an aquifer the general equation of flow (3.14-19) applies. Also, as will be shown below a similar equation in terms of the mean hydraulic head is capable of defining the shape of the piezometric surface as a whole.

The mean hydraulic head is defined as

$$\bar{h} = \frac{1}{b} \int_0^b h \, dz \qquad (5.1\text{-}1)$$

To obtain the equation for the piezometric surface we begin by integrating the general flow equation with respect to the coordinate "z" between limits

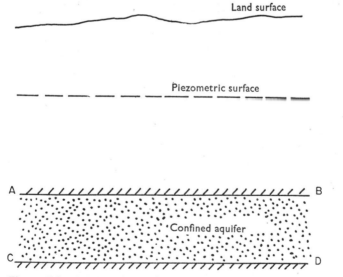

Figure 5.1 A horizontal, uniformly-thick, confined aquifer.

"0" and "b" (thickness of the aquifer), that is,

$$\int_0^b \frac{\partial^2 h}{\partial x^2}\, dz + \int_0^b \frac{\partial^2 h}{\partial y^2}\, dz + \int_0^b \frac{\partial^2 h}{\partial z^2}\, dz = \frac{S_s}{K}\int_0^b \frac{\partial h}{\partial t}\, dz$$

This equation can be reduced to the form

$$\frac{\partial^2}{\partial x^2}\int_0^b h\, dz + \frac{\partial^2}{\partial y^2}\int_0^b h\, dz + \int_0^b \frac{\partial}{\partial z}\left(\frac{\partial h}{\partial z}\right) dz = \frac{S_s}{K}\frac{\partial}{\partial t}\int_0^b h\, dz \qquad (5.1\text{-}2)$$

and in terms of the mean hydraulic head $\bar h$ equation (5.1-2) can be written as

$$\frac{\partial^2}{\partial x^2}(b\bar h) + \frac{\partial^2}{\partial y^2}(b\bar h) + \frac{\partial h}{\partial z}\bigg|_b - \frac{\partial h}{\partial z}\bigg|_0 = \frac{S_s}{K}\frac{\partial}{\partial t}(b\bar h) \qquad (5.1\text{-}3)$$

Since at $z = 0$ and at $z = b$ the velocity is zero, it follows that at these points

$$\frac{\partial h}{\partial z} = -\frac{v_z}{K} = 0$$

and equation (5.1-3) reduces to

$$\frac{\partial^2 \bar h}{\partial x^2} + \frac{\partial^2 \bar h}{\partial y^2} = \frac{S_s}{K}\frac{\partial \bar h}{\partial t} \qquad (5.1\text{-}4)$$

or

$$\frac{\partial^2 \bar h}{\partial x^2} + \frac{\partial^2 \bar h}{\partial y^2} = \frac{S}{T}\frac{\partial \bar h}{\partial t} \qquad (5.1\text{-}5)$$

where S is the storage coefficient and T is the transmissivity.

5.2   Equation of Flow for a Leaky Horizontal Aquifer of Uniform Thickness

In the case of a leaky aquifer where the face AB (see Figure 5.2) is semi-permeable, the aquifer receives a certain amount of water from the leaky layer located above AB. The thickness of this layer will be taken as b' and its hydraulic conductivity as K'. Thus,

$$v_z|_b = K'\frac{h - h_1}{b'} \qquad (5.2\text{-}1)$$

where h_1 is the distance from CD to the water level. Consequently, combining (5.1-3) and (5.2-1) we obtain

$$\frac{\partial^2}{\partial x^2}(b\bar h) + \frac{\partial^2}{\partial y^2}(b\bar h) + \frac{K'}{K}\frac{h_1 - \bar h}{b'} = \frac{S_s}{K}\frac{\partial}{\partial t}(b\bar h) \qquad (5.2\text{-}2)$$

which reduces to

$$\frac{\partial^2 \bar{h}}{\partial x^2} + \frac{\partial^2 \bar{h}}{\partial y^2} + \frac{h_1 - \bar{h}}{B^2} = \frac{S}{T}\frac{\partial \bar{h}}{\partial t} \tag{5.2-3}$$

where $B = \sqrt{T/(K'/b')}$ = Leakage factor and K'/b' is the Leakage coefficient or Leakance. This coefficient varies depending on the formations involved. Most reported values range from 10^{-7} to 10^{-10} sec^{-1}.

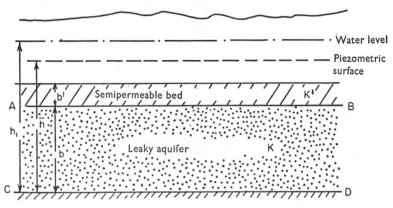

Figure 5.2 A leaky horizontal, uniformly-thick aquifer.

5.3 Equation of Flow for a Confined Inclined Aquifer of Uniform Thickness

Consider now a fully confined aquifer inclined an angle θ with respect to the horizontal as shown on Figure 5.3. To obtain the differential equation representing this case we proceed as done in Section 5.1 to average the general flow equation in such a way that one obtains an equation in terms of the average head $\bar{h}$. Thus,

$$\int_{z_0}^{z_1}\frac{\partial^2 h}{\partial x^2}\,dz + \int_{z_0}^{z_1}\frac{\partial^2 h}{\partial y^2}\,dz + \int_{z_0}^{z_1}\frac{\partial^2 h}{\partial z^2}\,dz = \frac{S}{T}\int_{z_0}^{z_1}\frac{\partial h}{\partial t}\,dz \tag{5.3-1}$$

where z_0 and z_1 are defined in Figure 5.3.

Now, one must apply Leibniz's rule and recall that $v_z = mv_x$ along the aquifer boundaries. Therefore,

$$\frac{1}{b}\left(\frac{\partial}{\partial x}\int_{z_0}^{z_1}\frac{\partial h}{\partial x}\,dz + \frac{\partial}{\partial y}\int_{z_0}^{z_1}\frac{\partial h}{\partial y}\,dz\right) = \frac{S}{T}\frac{\partial \bar{h}}{\partial t} \tag{5.3-2}$$

and this equation reduces to

$$\frac{\partial^2 \bar{h}}{\partial x^2} + \frac{m}{b}\left(\frac{\partial h}{\partial x}\right)_{z1} - \frac{m}{b}\left(\frac{\partial h}{\partial x}\right)_{-0} + \frac{\partial^2 \bar{h}}{\partial y^2} = \frac{S}{T}\frac{\partial \bar{h}}{\partial t} \tag{5.3-3}$$

which gives the average head for a confined aquifer that is inclined an angle θ with respect to the horizontal and that possesses a uniform thickness.

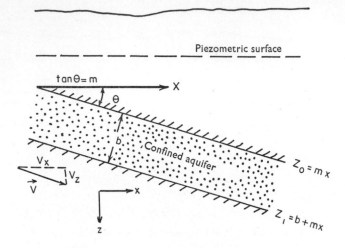

Figure 5.3 An inclined confined aquifer.

5.4 Equation of Flow for an Aquifer of Variable Thickness

Consider a confined aquifer as depicted on Figure 5.4 where the thickness b is a variable function of the space coordinates x and y. Thus,

$$b = f(x, y)$$

The first step that we must take in order to derive the differential equation for this case consists, as before, in averaging the flow equation, that is,

$$\int_0^b \frac{\partial^2 h}{\partial x^2}\,dz + \int_0^b \frac{\partial^2 h}{\partial y^2}\,dz + \int_0^b \frac{\partial^2 h}{\partial z^2}\,dz = \frac{S_s}{K}\int_0^b \frac{\partial h}{\partial t}\,dz \tag{5.4-1}$$

This equation reduces to

$$\frac{1}{b}\int_0^b \frac{\partial^2 h}{\partial x^2}\,dz + \frac{1}{b}\int_0^b \frac{\partial^2 h}{\partial y^2}\,dz = \frac{S}{Tb}\int_0^b \frac{\partial h}{\partial t}\,dz \tag{5.4-2}$$

But, it can be shown that

$$\frac{1}{b} \int_0^b \frac{\partial^2 h}{\partial x^2} \, dz = \frac{\partial^2 \bar{h}}{\partial x^2} + \frac{1}{b} \frac{\partial b}{\partial x} \frac{\partial h}{\partial x} \bigg|_{f(x,y)} \qquad (5.4\text{-}3)$$

and also

$$\frac{1}{b} \int_0^b \frac{\partial^2 h}{\partial y^2} \, dz = \frac{\partial^2 \bar{h}}{\partial y^2} + \frac{1}{b} \frac{\partial b}{\partial y} \frac{\partial h}{\partial y} \bigg|_{f(x,y)} \qquad (5.4\text{-}4)$$

Therefore, combining (5.4-2), (5.4-3), and (5.4-4) one finally obtains

$$\frac{\partial^2 \bar{h}}{\partial x^2} + \frac{\partial^2 \bar{h}}{\partial y^2} + \frac{1}{b} \left(\frac{\partial b}{\partial x} \frac{\partial h}{\partial x} \bigg|_{f(x,y)} + \frac{\partial b}{\partial y} \frac{\partial h}{\partial y} \bigg|_{f(x,y)} \right) = \frac{S}{T} \frac{\partial \bar{h}}{\partial t} \qquad (5.4\text{-}5)$$

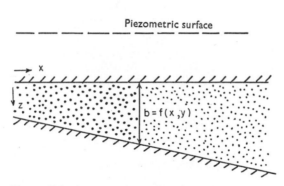

Figure 5.4 A confined aquifer of variable thickness.

5.5 Equation of Flow for an Unconfined Aquifer

Unconfined or water table aquifers (Figure 5.5) are those whose upper boundary is the water table which is a surface at atmospheric pressure (see Section 3.6). This type of aquifer is more difficult to treat than the confined case because of the difficulty of handling the water table as a free surface. Therefore, for most practical purposes, unconfined aquifer flow is treated in an approximate manner using the Dupuit conditions as will be shown in the following section. Nevertheless, more exact equations can be obtained both in terms of the potential function and in terms of the hydraulic head. Also, equations exist that are capable of including the effect of recharge on an unconfined aquifer.

At any point in an unconfined aquifer the velocity potential (see Chapter

4) is given by

$$\phi(x, y, z, t) = \frac{p}{\gamma} + z \tag{5.5-1}$$

where p is the pressure, $\gamma = \rho g$, and z is the head elevation.

Starting with equation (5.5-1) it can be shown (proofs left to the reader) that

$$\left(\frac{\partial \phi}{\partial x}\right)^2 + \left(\frac{\partial \phi}{\partial y}\right)^2 + \left(\frac{\partial \phi}{\partial z}\right)^2 - \frac{\partial \phi}{\partial z} = \frac{R}{K}\frac{\partial \phi}{\partial t} \tag{5.5-2}$$

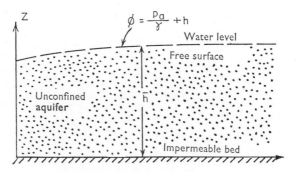

Figure 5.5   An unconfined aquifer.

and

$$\left(\frac{\partial h}{\partial x}\right)^2 + \left(\frac{\partial h}{\partial y}\right)^2 + \left(\frac{\partial h}{\partial z}\right)^2 - \frac{\partial h}{\partial z} = \frac{R}{K}\frac{\partial h}{\partial t} \tag{5.5-3}$$

where R is the specific yield (see Chapter 3), that is, that quantity of fluid which will be released by a unit volume of the aquiferous formation as a result of gravitational load.

Equations (5.5-2) and (5.5-3) give the potential and head variations (respectively) for an unconfined aquifer.

When recharge directly into the aquifer is important the approximate equation that should be used is

$$\nabla^2 h^2 = -\frac{W}{2K} \tag{5.5-4}$$

where W is the rate of recharge per unit area.

5.6 Approximate Equation of Flow for an Unconfined Aquifer

To approximate the treatment of unconfined flow by a simple equation, Dupuit (1863) in his *Etudes Theoriques et Pratiques sur le Mouvement des Eaux* indicated that two assumptions must be made. These are:

1. The fluid's velocity is proportional to the tangent of the hydraulic gradient, and
2. the flow is uniform through a vertical cross-section, that is, for small inclinations of the water table the streamlines can be taken to be horizontal.

To obtain an approximate equation for unconfined flow using these assumptions one can begin by noting that the equation of continuity would require and imply for the case of steady flow in an unconfined aquifer that

$$\frac{\partial}{\partial x}(\bar{h}v_x) + \frac{\partial}{\partial y}(\bar{h}v_y) = 0 \qquad (5.6\text{-}1)$$

where $\bar{h}$ is the average head above the impermeable bed.

Using Dupuit's assumptions equations for v_x and v_y can be written such that

$$v_x = -K\frac{\partial \bar{h}}{\partial x}, \qquad v_y = -K\frac{\partial \bar{h}}{\partial y} \qquad (5.6\text{-}2)$$

Then, combining (5.6-1) and (5.6-2) one obtains that

$$\frac{\partial^2 \bar{h}^2}{\partial x^2} + \frac{\partial^2 \bar{h}^2}{\partial y^2} = 0 \qquad (5.6\text{-}3)$$

This last equation approximately represents steady flow of water in an unconfined aquifer.

5.7 Approximate Equation of Flow for an Inclined Unconfined Aquifer

In Section 5.3 the case of an inclined confined aquifer was considered, and we obtained that equation (5.3-3) mathematically represents the situation. The equivalent case for an unconfined aquifer (see Figure 5.6) can be similarly handled and the reader should by now be capable of deriving on the basis of strict mathematics the required differential equation.

It is also possible to obtain the required equation using simple analogies. It has already been shown that while h is a potential function for confined flow, h^2 is the potential function for unconfined flow. Also in an unconfined aquifer one does not have a clearcut thickness b as in the case of confined aquifers, but must work in terms of the saturated thickness $\bar{z}$, where $\bar{z}$ in the case being considered here is the water level measured from the inclined impermeable bed and is mathematically equal to $h - f$ (see Figure 5.6). Thus,

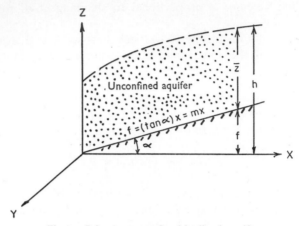

Figure 5.6 An unconfined inclined aquifer.

for an inclined unconfined aquifer, equation (5.3-3) becomes in terms of $\bar{z}$

$$\frac{\partial^2 \bar{z}^2}{\partial x^2} + \frac{m}{\bar{z}} \frac{\partial \bar{z}^2}{\partial x} + \frac{\partial^2 \bar{z}^2}{\partial y^2} = \frac{S_s}{K\bar{z}} \frac{\partial \bar{z}^2}{\partial t} \tag{5.7-1}$$

or in terms of an average depth $\bar{d}$ equation (5.7-1) can be written as

$$\frac{\partial^2 \bar{z}^2}{\partial x^2} + \frac{m}{\bar{d}} \frac{\partial \bar{z}^2}{\partial x} + \frac{\partial^2 \bar{z}^2}{\partial y^2} = \frac{S_s}{K\bar{d}} \frac{\partial \bar{z}^2}{\partial t} \tag{5.7-2}$$

5.8 Solution of Leaky Aquifer Problems Using a Principle of Correspondence

The treatment of leaky aquifers is much more complicated than that of equivalent confined aquifers. Jacob[4] and Hantush[3] pioneered the theory of

---

[4] Jacob, C. E. (1946) *Radial Flow in a Leaky Aquifer*, Transactions American Geophysical Union, volume 27.

leaky aquifers. Herrera and Figueroa have recently modified this theory and simplified it considerably.

Herrera and Figueroa[5] developed a correspondence principle that states that "to every leaky aquifer with appreciable storage capacity on the leaky layer, corresponds a certain fully confined aquifer." This correspondence is such that for slow variations in head, the behavior of the leaky aquifer can be predicted by a simple variable change when the behavior of the corresponding confined aquifer is known.

As done in the previous sections of this chapter, the differential equations for the leaky aquifer case using the principle of correspondence will now be discussed. For an easier understanding of the subject, three cases will be considered.

Case 1. The leaky (semipermeable) layers are above and below two other aquifers where the hydraulic head remains constant. In this case the following system of equations must be solved:

1. Upper Leaky Bed:

$$\frac{\partial^2 s_1}{\partial z^2} = \frac{1}{\nu_1} \frac{\partial s_1}{\partial t} \tag{5.8-1}$$

$$s_1(x, y, z, 0) = 0 \tag{5.8-2}$$

$$s_1(x, y, b_1, t) = 0 \tag{5.8-3}$$

$$s_1(x, y, 0, t) = s(x, y, t) \tag{5.8-4}$$

2. Main Aquifer:

$$\frac{\partial^2 s}{\partial x^2} + \frac{\partial^2 s}{\partial y^2} + \frac{K_1}{T} \frac{\partial s_1}{\partial z}(x, y, 0, t) - \frac{K_2}{T} \frac{\partial s_2}{\partial z}(x, y, 0, t) = \frac{1}{\nu} \frac{\partial s}{\partial t} \tag{5.8-5}$$

$$s(x, y, 0) = 0 \tag{5.8-6}$$

3. Lower Leaky Bed:

$$\frac{\partial^2 s_2}{\partial z^2} = \frac{1}{\nu_2} \frac{\partial s_2}{\partial t} \tag{5.8-7}$$

$$s_2(x, y, z, 0) = 0 \tag{5.8-8}$$

$$s_2(x, y, 0, t) = s(x, y, t) \tag{5.8-9}$$

$$s_2(x, y, -b_2, t) = 0 \tag{5.8-10}$$

[5] Herrera, I. and Figueroa, G. E. (1968) *A Principle of Correspondence for the Theory of Leaky Aquifers*, Geophysical Institute, University of Mexico (UNAM), Mexico, D.F., publication 1001.

where the numbers 1 and 2 refer to the upper and lower leaky beds respectively and

$$v = \frac{T}{S} = \text{Hydraulic Diffusivity}$$

Case 2. The leaky beds are below and above two impermeable strata. In this case the system of equations to be solved is the same as in the previous case except for a substitution of equations (5.8-3) and (5.8-10) respectively by the equations

$$\frac{\partial s_1}{\partial z}(x, y, b_1, t) = 0 \qquad (5.8\text{-}11)$$

and

$$\frac{\partial s_2}{\partial z}(x, y, -b_2, t) = 0 \qquad (5.8\text{-}12)$$

Case 3. The lower leaky bed rests over an impermeable formation while the upper leaky bed is below an aquifer of constant hydraulic head. To treat this case one solves the system in Case 1 but substitutes (5.8-10) by (5.8-12).

It is important to mention that in the Herrera-Figueroa model considered above any influence exerted by a leaky layer on a neighboring formation is exerted through the main aquifer. Also, the influence that the leaky layers exert on the main aquifer is felt in the aquifer's yield which is characterized by the functions

$$\frac{\partial s_1}{\partial z}(x, y, 0, t) \quad \text{and} \quad \frac{\partial s_2}{\partial z}(x, y, 0, t)$$

Consequently, and on the basis of these observations, Herrera and Figueroa were able to reduce the complex systems of simultaneous equations discussed above to an integro-differential equation where the integral term represents the memory of the aquifer to prior events.

In a subsequent investigation Herrera[6] has extended his theory to include the treatment of multiple leaky aquifers by means of a correspondence principle. Since this case is rather complicated it is beyond the scope of this book.

---

[6] Herrera, I. (1969) *Theory of Multiple Leaky Aquifers*, Geophysical Institute of Mexico, pub. 1014.

PART 3

WELL HYDRAULICS

CHAPTER 6

Steady Flow

In the previous chapter, the flow equations for various types of aquifers have been derived. These equations allow the representation of a multitude of physical problems in groundwater hydrology. As has already been pointed out in Chapter 3, these problems can be grouped into steady and unsteady flow. In this chapter the steady flow regime is analyzed, that is, we will concern ourselves here with problems where the conditions

$$\frac{\partial p}{\partial t} = 0, \qquad \frac{\partial h}{\partial t} = 0$$

are satisfied. Problems where these conditions are not met will be treated in the following chapter.

Steady state flow analysis is useful in that it allows an understanding of flow conditions in an aquifer and it permits a representation of piezometric surface or head contours.

In applying the techniques discussed in this chapter to specific situations, the reader must be sure that an equilibrium state has been achieved. If this is not the case the methods discussed here do not apply.

6.1 Linear Flow in a Simple Confined Aquifer

The simplest situation which can be considered is that of a confined, horizontal aquifer that is fully saturated with an incompressible fluid such as water. If the aquifer is homogeneous and isotropic, then, as previously shown for steady state conditions the flow equation is

$$\frac{\partial^2 h}{\partial x^2} + \frac{\partial^2 h}{\partial y^2} + \frac{\partial^2 h}{\partial z^2} = 0 \qquad (6.1\text{-}1)$$

To simplify matters further consider the flow as being unidirectional (linear) as shown on Figure 6.1. Then, equation (6.1-1) reduces to

$$\frac{\partial^2 h}{\partial x^2} = 0 \tag{6.1-2}$$

Two boundary conditions with respect to the coordinate x are required to solve equation (6.1-2). These conditions are

$$h(0) = h_1 \quad \text{and} \quad h(l) = h_2$$

The solution to equation (6.1-2) is easily found to be

$$h = c_1 x + c_2 \tag{6.1-3}$$

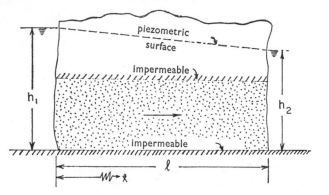

Figure 6.1   Unidirectional flow in a confined aquifer.

and applying the above boundary conditions one obtains that

$$h = h_1 - \frac{h_1 - h_2}{l} x \tag{6.1-4}$$

This last equation gives the piezometric head surface at all points x_i from one extreme of the aquifer to the other. Also, if to equation (6.1-4) one applies Darcy's law, the total discharge Q through the aquifer can easily be obtained.

6.2   Linear Flow in a Simple Unconfined Aquifer

In Section 5.6 of the previous chapter it was shown that in order to solve the problem of flow in an unconfined aquifer it was convenient to assume that the fluid's velocity is proportional to the tangent of the hydraulic gradient and that for small inclinations of the water table the streamlines are horizontal. These are Dupuit's assumptions and they are important because of the difficulties of handling the water table by other means. It was also shown in the previous chapter that for unconfined aquifers the function h^2 is a potential function just as h is a potential function for the case of a confined aquifer. Thus, for the case of steady unidirectional flow in an unconfined, horizontal, homogeneous aquifer the flow is represented by

$$\frac{\partial h^2}{\partial x^2} = 0 \qquad (6.2\text{-}1)$$

Equation (6.2-1) is analogous to equation (6.1-2) except that, as pointed out above, the case considered here involves h^2.

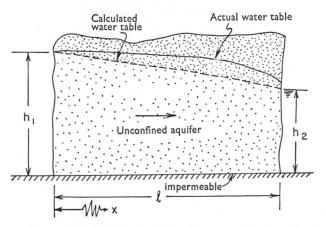

Figure 6.2 Unidirectional flow in an unconfined aquifer.

For the case under study (see Figure 6.2) the boundary conditions are

$$h(0) = h_1$$
$$h(l) = h_2 \qquad (6.2\text{-}2)$$

The solution to equation (6.2-1) is similar to the solution to equation (6.1-2), that is

$$h^2 = c_1 x + c_2 \qquad (6.2\text{-}3)$$

Substituting the above boundary conditions (6.2-2) into equation (6.2-3) one obtains

$$h^2 = \frac{h_2^2 - h_1^2}{l} x + h_1^2 \qquad (6.2\text{-}4)$$

In the same way as for a confined aquifer the discharge Q can be obtained by a direct application of Darcy's law.

Equation (6.2-4) represents the shape of the water table for an unconfined aquifer. Mathematically speaking, this is the equation of a parabola usually known as the Dupuit parabola. On the other hand, for the analogous case of a confined aquifer considered in Section 6.1 the piezometric surface is represented by a straight line.

In Figure 6.2 two curves are shown as representing the water table and the lower one is the Dupuit parabola as calculated above. The actual water table is the upper curve. Notice that the true free surface (the actual water table) must be horizontal at $x = 0$ and approach $x = l$ tangentially. The reason why the true water table must be horizontal at $x = 0$ is that this face is an equipotential. Similarly, the reason for the tangential approach at $x = l$ is that the true free surface cannot meet at $x = l$ at the point $h = h_2$ but must tangentially terminate at a point above h_2 and merge into a *surface of seepage* that extends from that point to $h = h_2$. The surface of seepage is not a streamline, however, it is exposed to a uniform pressure. For further details concerning free surfaces the reader can refer to Morris Muskat's treatment of gravity flow systems.[1]

6.3 Linear Flow in a Leaky Aquifer

In the previous chapter (see Section 5.2) the differential equation that governs the flow in a leaky, horizontal, uniformly-thick aquifer was derived. For the case of steady unidirectional flow this equation can be written in the form

$$\frac{\partial^2 h}{\partial x^2} + \frac{h_0 - h}{B^2} = 0 \tag{6.3-1}$$

where B is the leakage factor and is defined by the equation

$$B = \sqrt{\frac{Tb'}{K'}} \tag{6.3-2}$$

where T is the aquifer's transmissivity and b' and K' are respectively the average thickness and hydraulic conductivity of the leaky bed (see Figure 6.3).

As before we can use as boundary conditions

$$h(0) = h_1$$
$$h(l) = h_2 \tag{6.3-3}$$

To obtain a solution to our problem we must note that equation (6.3-1) is

---

[1] Muskat, M. (1946) *The Flow of Homogeneous Fluids Through Porous Media*, J. W. Edwards, Inc., Ann Arbor, Michigan, USA.

of the form

$$\frac{\partial^2 h}{\partial x^2} + c_1 h = c_2 \qquad (6.3\text{-}4)$$

where $c_1 = -1/B^2$ and $c_2 = -h_0/B^2$ are constants. The solution to an equation of this form can be easily found in any book on elementary differential equations and can easily be expressed in terms of the hyperbolic sine and cosine. Thus, the solution to (6.3-1) is

$$h = C_1 \cosh x/B + C_2 \sinh x/B + h_0 \qquad (6.3\text{-}5)$$

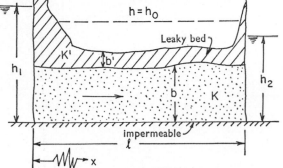

Figure 6.3 Unidirectional flow in a leaky aquifer.

The constants C_1 and C_2 can be evaluated by substituting the boundary conditions (6.3-3) into (6.3-5). Therefore, one obtains that for such a leaky aquifer under steady state conditions

$$h = h_0 + (h_1 - h_0)\cosh x/B + \left(\frac{h_2 - h_0 - (h_1 - h_0)\cosh l/B}{\sinh l/B} \right) \sinh x/B$$

$$(6.3\text{-}6)$$

This equation gives values that are in any case intermediate between the fully confined and unconfined cases. Values will approach one or the other case depending on the value of B, that is, on the extent of leakage.

6.4 Linear Flow in an Unconfined Inclined Aquifer

In Section 5.7 of the previous chapter the differential equation that governs the flow in an unconfined inclined aquifer was derived. The case of unidirectional steady flow in such an aquifer is depicted in Figure 6.4 and will now be treated.

It is easier to handle this case by writing the differential equation in terms of head and the average depth of the water, $\bar{d}$, that is,

$$\bar{d} = \frac{h_1 + h_2}{2}$$

Thus,

$$\frac{\partial^2 h^2}{\partial x^2} - \frac{m}{\bar{d}} \frac{\partial h^2}{\partial x} = 0 \qquad (6.4\text{-}1)$$

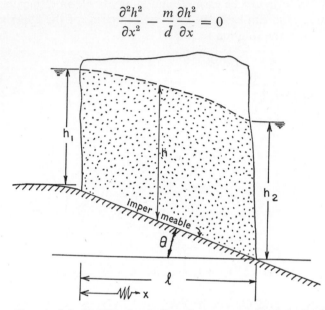

Figure 6.4 Unidirectional flow in an unconfined inclined aquifer.

As before, the boundary conditions are

$$h(0) = h_1 \qquad (6.4\text{-}2)$$

and

$$h(l) = h_2 \qquad (6.4\text{-}3)$$

To solve equation (6.4-1) we must first apply the substitutions

$$a = m/\bar{d} \quad \text{and} \quad u = h^2$$

Then,

$$\frac{\partial^2 u}{\partial x^2} = a \frac{\partial u}{\partial x} \qquad (6.4\text{-}4)$$

Now, if

$$w = \frac{\partial u}{\partial x}$$

then it follows that

$$\frac{\partial w}{\partial x} = aw \quad \text{and} \quad w = c \exp(ax)$$

Consequently, u can be calculated since

$$c \exp(ax) = \frac{\partial u}{\partial x}$$

and therefore

$$u = c_1 \exp(ax) + c_2$$

Thus,

$$h^2 = c_1 \exp(mx/\bar{d}) + c_2 \tag{6.4-5}$$

The constants c_1 and c_2 can be evaluated using the boundary conditions (6.4-2) and (6.4-3). Thus,

$$h_1^2 = c_1 + c_2 \tag{6.4-6}$$

and

$$h_2^2 = c_1 \exp(ml/\bar{d}) + c_2 \tag{6.4-7}$$

Solving equations (6.4-6) and (6.4-7) simultaneously we can obtain values for c_1 and c_2 that can then be substituted into (6.4-5) to obtain

$$h_1^2 - h^2 = \frac{h_1^2 - h_2^2}{(1 - \exp(ml/\bar{d}))} (1 - \exp(mx/\bar{d})) \tag{6.4-8}$$

This equation expresses the head changes as a function of the coordinate x for an unconfined inclined aquifer. Equation (6.4-8) can also be expressed using hyperbolic functions. This is left as an exercise for the reader.

6.5 Flow in an Aquifer of Variable Thickness

In Section 5.4 of the proceding chapter it was found that equation (5.4-5) governs the flow in an aquifer of variable thickness. Consider now the case of steady flow in a confined aquifer where the thickness varies only with respect to x (see Figure 6.5) according to the equation

$$b = b_0 \exp(-cx) \tag{6.5-1}$$

Then, equation (5.4-5) can be simplified to

$$\frac{\partial^2 h}{\partial x^2} - c \frac{\partial h}{\partial x} = 0 \tag{6.5-2}$$

The boundary conditions for the problem under consideration are the same as used before in this chapter, that is, at $x = 0$ the head $h = h_1$ and at

$x = l$ the head $h = h_2$. Also equation (6.5-2) is of the same form as equation (6.4-4). Therefore,

$$h = c_1 \exp(cx) + c_2 \qquad (6.5\text{-}3)$$

Finally, applying the boundary conditions above to equation (6.5-3) one obtains that the head in the aquifer considered here is given by the equation

$$h = h_1 - \frac{(h_1 - h_0)}{(1 - \exp(cl))}(1 - \exp(cx)) \qquad (6.5\text{-}4)$$

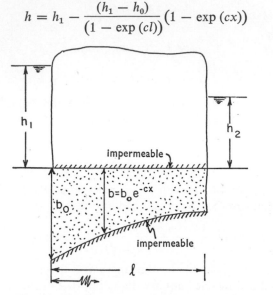

Figure 6.5 Flow in an aquifer of variable thickness.

6.6 Flow in a Faulted Aquifer

An important case in groundwater hydrology is that of an aquifer containing geologic faults or fractures. It is common to find aquifers where tectonic disturbances have caused breaks and/or dislocations. A portion of such an aquifer will be considered here (see Figure 6.6). The differential equation governing the steady flow in such an aquifer must, of course, be written in two dimensions since the fault here considered is perpendicular to the direction of fluid motion. Therefore, two boundary conditions with respect to each x and z are needed to solve the problem.

The differential equation to be solved is

$$\frac{\partial^2 h}{\partial x^2} + \frac{\partial^2 h}{\partial z^2} = 0 \qquad (6.6\text{-}1)$$

and the boundary conditions are

$$\left(\frac{\partial h}{\partial z}\right)_{x,b} = 0 \tag{6.6-2}$$

$$\left(\frac{\partial h}{\partial z}\right)_{x,0} = 0 \tag{6.6-3}$$

$$h(l, z) = h_1 \tag{6.6-4}$$

$$\frac{\partial h}{\partial x} = 0 \quad \text{when} \quad 0 < z < a \quad \text{and}$$
$$= q/K(b - a) \quad \text{when} \quad a < z < b \tag{6.6-5}$$

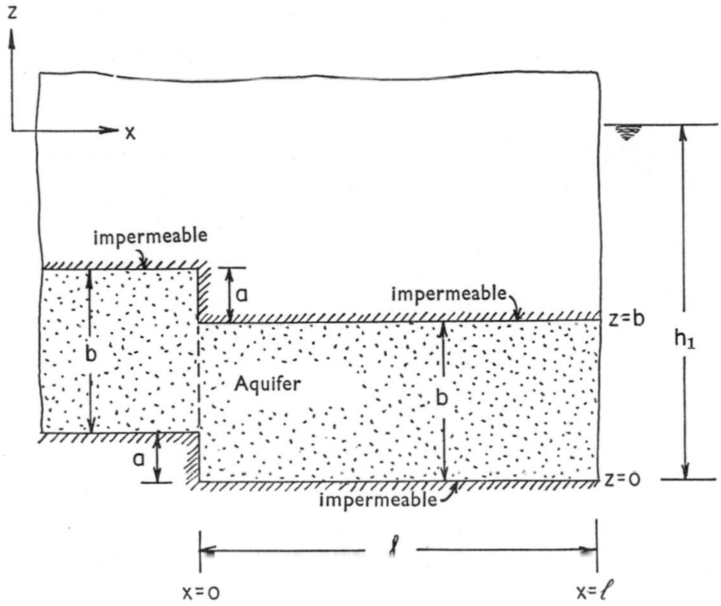

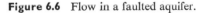

Figure 6.6 Flow in a faulted aquifer.

Equation (6.6-1) can be solved using the method of separation of variables. Thus, we assume that $h = h_1 + X(x)Z(z)$ and therefore (6.6-1) becomes

$$\frac{X''(x)}{X} = -\frac{Z''(z)}{Z} = p^2 \tag{6.6-6}$$

Solving equation (6.6-6) the following possible solutions are obtained

$$X = \sinh p(l - x) \quad \text{or} \quad X = \cosh p(l - x) \tag{6.6-7}$$

and

$$Z = \sin pz \quad \text{or} \quad Z = \cos pz \tag{6.6-8}$$

Using the boundary conditions (6.6-3) and (6.6-4) one can notice that $X = \cosh p(l - x)$ and $Z = \sin pz$ do not satisfy these conditions. Therefore,

$$h = h_1 + C \sinh p(l - x) \cos pz \tag{6.6-9}$$

Now, applying condition (6.6-2) to equation (6.6-9) one obtains that $\sin pb = 0$ and therefore $pb = n\pi$.

Consequently equation (6.6-9) can be generalized and written in the form

$$h = h_1 + \sum_{n=0}^{\infty} A_n \sinh \frac{n\pi}{b} (l - x) \cos \frac{n\pi z}{b} \tag{6.6-10}$$

As one can easily see equation (6.6-10) is a type of Fourier expansion and using the Fourier method A_n can be evaluated with the aid of equation (6.6-5). In this way one finally obtains that

$$h = h_1 + \frac{q}{Kb} (x - l)$$

$$+ \frac{2qb}{\pi^2 K(b - a)} \sum_{n=1}^{\infty} \left(\frac{\sin \dfrac{n\pi a}{b}}{n^2 \cosh \dfrac{n\pi l}{b}} \right) \sinh \frac{n\pi}{b} (l - x) \cos \frac{n\pi z}{b} \tag{6.6-11}$$

6.7 Radial Flow into a Well Fully Penetrating a Confined Aquifer

Up to now all of the problems that have been solved involved the use of cartesian coordinates. In certain instances it is more practical to use polar, cylindrical or spherical coordinates. One such case is the classical problem of a well fully penetrating an extensive confined aquifer in a circular island (see Figure 6.7). In this case due to the existence of radial symmetry it is best to use Laplace's equation in polar coordinates (see Section 3.20). Thus, our problem consists in solving the equation

$$\frac{\partial^2 h}{\partial r^2} + \frac{1}{r} \frac{\partial h}{\partial r} = 0 \tag{6.7-1}$$

Also assume that at the well face

$$h = h_w \tag{6.7-2}$$

and

$$\lim_{r \to 0} r \frac{\partial h}{\partial r} = \frac{Q}{2\pi T} \qquad (6.7\text{-}3)$$

From equation (6.7-1) and condition (6.7-3) it follows that

See Bear,
pp. 304- 305

$$r \frac{\partial h}{\partial r} = \frac{Q}{2\pi T} \rightarrow r \frac{d^2 h}{d r^2} + \frac{d h}{d r} \doteq 0$$

Q must be the volume
rate — See Eqn 6.7-8 for
dimensional
analysis

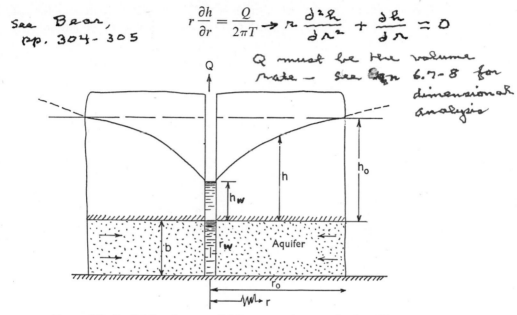

Figure 6.7   Radial flow into a well fully penetrating a confined aquifer
in a circular island.

and consequently

$$h = \frac{Q}{2\pi T} \ln r + c \qquad (6.7\text{-}4)$$

Using equation (6.7-2) the constant c can be evaluated and we obtain that

$$h - h_w = \frac{Q}{2\pi T} \ln \frac{r}{r_w} \qquad (6.7\text{-}5)$$

This last equation indicates that h increases indefinitely as r increases. Of course, we know that in reality $h = h_0$ when $r = r_0$, where r_0 is the distance beyond which the influence of the well is not felt. To include this condition we let $h = h_0$ at $r = r_0$ be then substituted into equation (6.7-5) and obtain

$$h_0 - h_w = \frac{Q}{2\pi T} \ln \frac{r_0}{r_w} \qquad (6.7\text{-}6)$$

and from equations (6.7-5) and (6.7-6) follows that

$$h - h_w = (h_0 - h_w) \frac{\ln (r/r_w)}{\ln (r_0/r_w)} \qquad (6.7\text{-}7)$$

This last equation shows that the head varies linearly with the logarithm of the distance from the well. Equation (6.7-7) physically represents the case shown on Figure 6.8. Equation (6.7-6) is known as Thiem's equation.[2]

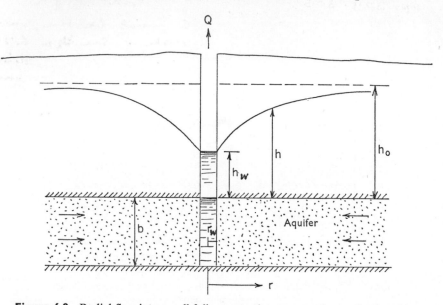

Figure 6.8   Radial flow into a well fully penetrating an extensive confined aquifer.

Thiem's equation is useful in calculating the hydraulic conductivity of a confined aquifer. For example, if we have two observation wells at distances r_1 and r_2 from the producing well and we observe heads in the observation wells of h_1 and h_2 respectively then by Thiem's equation

$$K = \frac{Q}{2\pi b(h_2 - h_1)} \ln \frac{r_2}{r_1} \qquad (6.7\text{-}8)$$

This last equation requires the following conditions:

(a) the aquifer must be homogeneous,
(b) a condition of equilibrium must exist,
(c) the aquifer is infinite in areal extent,

---

[2] Thiem, G. (1906) *Hydrologische Methoden*, Gebhardt, Leipzig, Germany.

(d) the well fully penetrates the aquifer, and

(e) the flow is laminar.

Even though these conditions are somewhat restrictive the Thiem equation is a good technique for calculating the hydraulic conductivity of a confined aquifer.

6.8 Radial Flow into a Well Fully Penetrating an Unconfined Aquifer

The equation of groundwater flow in an unconfined aquifer can be written in cylindrical coordinates for the case of radial symmetry as

$$\frac{\partial^2 h^2}{\partial r^2} + \frac{1}{r}\frac{\partial h^2}{\partial r} = 0 \tag{6.8-1}$$

Consider now such a horizontal unconfined aquifer (see Figure 6.9) where the head at the well radius r_w is

$$h(r_w) = h_w \tag{6.8-2}$$

and the head at the radius of influence is

$$h(r_0) = h_0 \tag{6.8-3}$$

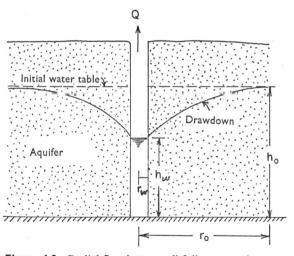

Figure 6.9 Radial flow into a well fully penetrating an unconfined aquifer.

As in Section 6.7 the solution to (6.8-1) is found to be

$$h^2 = c_1 \ln r + c_2$$

where c_1 and c_2 are constants.

Then, using equations (6.8-2) and (6.8-3) one obtains that

$$h^2 = h_0^{\,2} - \frac{h_0^{\,2} - h_w^{\,2}}{\ln (r_w/r_0)} \ln \frac{r}{r_0} \tag{6.8-4}$$

If instead of using condition (6.8-2) one uses

$$\lim_{r \to 0} r \frac{\partial h}{\partial r} = \frac{Q}{2\pi K h}$$

then one obtains that

$$Q = \pi K \frac{h_0^{\,2} - h^2}{\ln (r_0/r)} \tag{6.8-5}$$

This last equation is known as *Dupuit's Equation*.

The values of the radius of influence r_0 are in most cases rather arbitrarily defined and all they can hope to be is approximations. Generally r_0 varies between 150 and 300 meters.

Just as in the preceding section the hydraulic conductivity K can be determined using two observation wells, that is,

$$K = \frac{Q \ln r_1/r_2}{\pi(h_1^{\,2} - h_2^{\,2})} \tag{6.8-6}$$

6.9 Radial Flow Into a Well Fully Penetrating a Leaky Aquifer

Consider now the flow into a well fully penetrating a leaky aquifer (see Figure 6.10). In Section 5.2 the equation of groundwater flow in such an aquifer was given. This equation can be written in polar coordinates in the form

$$\frac{\partial^2 h}{\partial r^2} + \frac{1}{r} \frac{\partial h}{\partial r} + \frac{h_0 - h}{B^2} = 0 \tag{6.9-1}$$

where B is the leakage factor that was defined in Section 5.2.

If the aquifer is considered infinite in areal extent then we can use the constraint of zero drawdown infinitely far from the well, that is,

$$h(\infty) = h_0 \tag{6.9-2}$$

As an additional boundary condition we can say that at the well

$$\lim_{r \to 0} r \frac{\partial h}{\partial r} = \frac{Q}{2\pi T} \tag{6.9-3}$$

Equation (6.9-1) is the modified Bessel Equation. The solution to this equation can then be expressed in terms of the Bessel functions I and K of

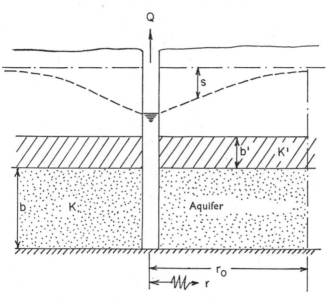

Figure 6.10 Radial flow into a well fully penetrating a leaky aquifer.

order zero and argument r/B as

$$h = h_0 + c_1 I_0(r/B) + c_2 K_0(r/B) \tag{6.9-4}$$

The function $I_0(\xi)$ is a real function called the modified Bessel function of the first kind of order zero. For any order n this function is defined by the equation

$$I_n(\xi) = (\sqrt{-1})^{-n} J_n(i\xi) \qquad i = \sqrt{-1}$$

where J_n is the Bessel function of the first kind of order n.

The function $K_0(\xi)$ is the modified Bessel function of the second kind of order zero. This function can be approximated for large values of the argument by the equation

$$K_0(\xi) = \sqrt{\frac{\pi}{2\xi}} \exp(-\xi)$$

and for small values of the argument by the formula

$$K_0(\xi) = -\ln\frac{\xi}{2}$$

The constants c_1 and c_2 can then be evaluated using the boundary conditions (6.9-2) and (6.9-3) and the definition of the above functions. Therefore one obtains that

$$h = h_0 + \frac{Q}{2\pi T} K_0(r/B) \qquad (6.9\text{-}5)$$

This last equation describes the head for the case of a well whose effect is felt up to infinity. If instead we want to consider an aquifer such that the effect of the well is only felt within a given radius of influence, then we must use instead of the boundary condition (6.9-2) the condition that at $r = r_0$ the head $h = h_0$. Then, instead of obtaining equation (6.9-5) our solution will be

$$h = h_0 + \frac{Q}{2\pi T} \left[K_0(r/B) - \frac{K_0(r_0/B)I_0(r/B)}{I_0(r_0/B)} \right] \qquad (6.9\text{-}6)$$

6.10 Flow Into a Well Penetrating Two Permeable Sands

Up to this point we have only considered cases involving a homogeneous single aquifer. In certain instances one encounters aquifers which are actually composed of two formations that differ in their transmissivity (see Figure 6.11). To solve such a case we must solve the flow equation in each of the zones while maintaining continuity of flow at the interface between the two aquiferous formations.

The method used here to solve the case of an aquifer composed of two formations can readily be extended to the case of n formations.

In the case under consideration let Q_1, s_1, and T_1 be the discharge, drawdown, and transmissivity of the lower aquifer while Q_2, s_2, and T_2 respectively refer to the upper aquifer. Then, in each zone the flow is described by the equation

$$\frac{\partial^2 s_i}{\partial r^2} + \frac{1}{r}\frac{\partial s_i}{\partial r} + \frac{\partial^2 s_i}{\partial z^2} = 0 \qquad (6.10\text{-}1)$$

where $i = 1$ or 2.

As boundary conditions one has that the planes $z = 0$ and $z = b$ represent impermeable boundaries and therefore no flow perpendicular to these planes can occur. Also, as seen before

$$\lim_{r \to 0} r \frac{\partial s_i}{\partial r} = -\frac{Q_i}{2\pi T_i}$$

and at the radius of influence the drawdown can be considered to be zero.

In addition to the boundary conditions one has to consider that at the interface

$$s_1(r, b_1) = s_2(r, b_1)$$

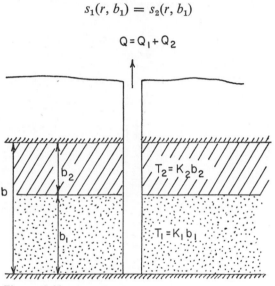

Figure 6.11 Flow into a well penetrating two permeable sands.

and

$$K_1 \frac{\partial s_1}{\partial z}(r, b_1) = K_2 \frac{\partial s_2}{\partial z}(r, b_1)$$

Equation (6.10-1) can be solved by the method of separation of variables and then the boundary and interface conditions can be applied to evaluate the resulting constants. Thus one obtains that

$$s_1 = \frac{Q_1}{2\pi T_1} \ln \frac{r_0}{r} + \frac{2K_1}{K_2} \left(\frac{Q_2}{\pi T_2} - \frac{Q_1}{\pi T_1} \right) \sum_{n=1}^{\infty} A_n \sinh (\alpha_n b_2) \cosh (\alpha_n z) J_0(\alpha_n r)$$

where the α_n are the roots of the equation $J_0(\alpha_n r_0) = 0$ and

$$A_n = \frac{1}{[\alpha_n r_0 J_1(\alpha_n r_0)]^2 [(1 + K_1/K_2) \sinh \alpha_n b - (1 - K_1/K_2) \sinh \alpha_n (b_1 - b_2)]}$$

The expression for s_2 is equally complicated. It is suggested that the reader exercise his mathematical ability in trying to find the equation for s_2.

6.11 Well Losses

In previous sections we have considered flow of water into a well for various typical situations. However, in all cases any effects due to the actual physical presence of the well were disregarded. These effects are mostly felt in the vicinity of the well and are termed *well losses*.

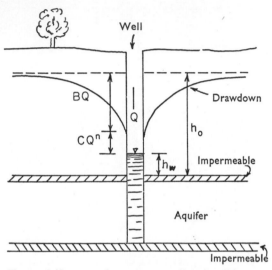

Figure 6.12 Drawdown near a well and well losses.

It was shown in Section 6.7 that the drawdown in a well, s_w, fully penetrating a confined aquifer can be obtained using the equation

$$s_w = h_0 - h_w = \frac{Q}{2\pi T} \ln \frac{r_0}{r_w} \qquad (6.11\text{-}1)$$

The above equation does not include well losses. Equation (6.11-1) can be applied without hesitation to those cases where the rate of pumping is low. However, if the well discharges are considerable well losses must be included as they may make up a large portion of the drawdown. These well losses may be greatly influenced by the occurrence of turbulent flow in the vicinity of the well.

Figure 6.12 shows the situation considered here. It can be seen from the figure that the actual head in the well is lower than that predicted by the drawdown curve as obtained by equation (6.11-1). The difference between this curve and the actual well head, h_w, is the well loss. Jacob[3] has suggested that in order to account for well losses equation (6.11-1) should be modified to

$$s_w = h_0 - h_w = BQ + CQ^2 \qquad (6.11\text{-}2)$$

where

$$B = \frac{\ln r_0/r_w}{2\pi T}$$

and C is a constant whose value depends on the well radius and the well's actual condition and deterioration.

Rorabaugh[4] has suggested that instead of using Jacob's equation above, one should use

$$s_w = h_0 - h_w = BQ + CQ^n \qquad (6.11\text{-}3)$$

where n varies (in most cases $n > 2$).

The constant C has the dimensions of $[T^2 L^{-5}]$ and is a rather important quantity which can be used as an indicator of how a well deteriorates. Walton[5] has devised some rules that help determine at least semiquantitatively the effectiveness of well development by measuring C. To measure C one uses the so called *Step Drawdown Tests* which will be discussed in the following chapter. Having so determined C Walton states that:

"1. The value of C of a properly developed and designed well is generally less than 5 sec²/ft⁵."

"2. Values of C between 5 and 10 sec²/ft⁵ indicate mild deterioration, and clogging is severe when C is greater than 10 sec²/ft⁵. It is difficult and sometimes impossible to restore the original capacity if the well-loss constant is greater than 40 sec²/ft⁵. Wells of diminished capacity can often be returned to near original capacity by one of several rehabilitation methods. The success of rehabilitation work can be appraised from the results of step-drawdown tests made prior to and after treatment."

If the discharge Q is divided by the drawdown we obtain the specific capacity. This quantity can be an index of well efficiency. For steady flow

$$\text{specific capacity} = \frac{1}{B + CQ^{n-1}} \qquad (6.11\text{-}4)$$

where $n = 2$ (Jacob) and n (Rorabaugh) is usually greater than 2 (average $\approx$ 2.5).

---

[3] Jacob, C. E. (1947) *Drawdown test to determine effective radius of artesian well*, Transactions ASCE, vol. 112, pp. 1047–1070.

[4] Rorabaugh, M. I. (1953) *Graphical and theoretical analysis of step-drawdown test of artesian well*, Proceedings ASCE, vol. 79, separate paper #362.

[5] Walton, W. C. (1962) *Selected analytical methods for well and aquifer evaluation*, Bulletin 49, Illinois State Water Survey.

Walton[5] also noted that if the specific capacity of a well is high then C tends to be high. He attributes this to increases in turbulence near the well that specially occur in formations where the transmissivity is low.

Lennox[6] extensively discusses the application of the Rorabaugh equation to the analysis of data from 18 well tests in the province of Alberta, Canada. Most of the wells considered by Lennox were drilled in low transmissivity formations and the Jacob equation could not be applied. In his particular case Lennox was able to obtain best results using Rorabaugh's equation with values for n as high as 3.5.

The above illustrates one of the uncertainties and problems involved with both the Jacob and Rorabaugh equation. Mogg[7] has pointed out this and other problems and concludes that what is actually needed is a critical revision of the step-drawdown test analysis. Mogg has also pointed out that in artesian aquifers in the vicinity of a pumping well a reduction in aquifer pressure would cause that a greater load be borne by the granular skeleton. If this skeleton has a low elastic modulus then compaction will set-in and cause a reduction in permeability in the zone of major pressure reduction, that is, in the vicinity of the well. This could cause a reduction in specific capacity that cannot be accounted for by turbulence as considered in both Jacob's and Rorabaugh's treatments. There are numerous other losses in the well, pipes, and screens that are not accounted for in equations (6.11-2) and (6.11-3). Thus, what is needed is a more adequate representation of the physical phenomena of flow near a well. Then, on the basis of this new representation a modified step-drawdown test could be designed.

6.12   Method of Images

A mathematical technique which is rather useful in solving hydrologic problems is the "method of images." This method consists in mirroring existing wells to obtain image wells in such a way that existing boundary conditions are satisfied.

To illustrate this technique let us find the drawdown in a point P produced by pumping a well at a distance r_1 from P when the well is very near the zone of recharge AB (see Figure 6.13).

It has already been shown in Section 6.7 that for the case of radial flow

---

[6] Lennox, D. H. (1966) *The analysis and application of the step-drawdown test*, Journal of the Hydraulics Division of ASCE, vol. 92, pp. 25–48.

[7] Mogg, J. L. (1968) *Step drawdown test needs critical review*, UOP Johnson Drillers Journal, vol. 40, n. 4, pp. 3–11.

into a well fully penetrating a confined aquifer

$$h = \frac{Q}{2\pi T} \ln r + c \qquad (6.12\text{-}1)$$

Then, since $h = h_0$ at $r = r_0$ (radius of influence), the quantity c can be evaluated and one obtains that the drawdown s is given by the formula

$$s = h_0 - h = \frac{Q}{2\pi T} \ln \frac{r_0}{r} \qquad (6.12\text{-}2)$$

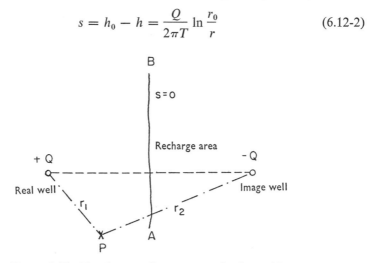

Figure 6.13   Flow into a well near a zone of recharge AB.

To determine the drawdown due to a well that is located near the zone of recharge we must apply the condition that $s = 0$ along the zone of recharge. To accomplish this it is necessary to mirror the real discharging well $+Q$ (using AB as a mirror) and construct an image recharging well $-Q$. The image and the real well will then be equidistant from the zone of recharge AB.

Now, if the superposition principle is applied using formula (6.12-2) the composite drawdown of both the real and image well can be found to be

$$s = \frac{Q}{2\pi T} \ln \frac{r_0}{r_1} - \frac{Q}{2\pi T} \ln \frac{r_0}{r_2} = \frac{Q}{2\pi T} \ln \frac{r_2}{r_1} \qquad (6.12\text{-}3)$$

It is easy to notice that formula (6.12-3) satisfies the condition that along AB the drawdown is zero since along this boundary $r_2 = r_1$.

In conclusion the method of images simply consists in mirroring existing wells about certain boundaries that must act as well-mirrors so that certain boundary conditions be satisfied. Once the system of real and image wells is complete the superposition principle is applied to obtain the combined total drawdown.

6.13 Flow into a Well Located in an Arbitrary Point in a Circular Aquifer

In a previous section, the problem of flow into a well located at the center of a circular island was treated. A more practical problem is that of a well eccentrically located with respect to a circular aquifer (see Figure 6.14). The method of images will be used to illustrate the solution process. In this particular case the boundary condition that must be satisfied is that the drawdown be zero at the aquifer's edge. Therefore, an image well $-Q$ must be placed outside the circle to balance the effect of the real well in such a way that along r_0 the drawdown $s = 0$.

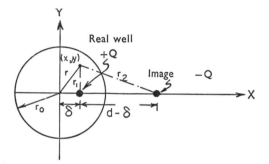

Figure 6.14   Flow into a well located in an arbitrary point in a circular aquifer.

The equation describing the drawdown due to the eccentric well is

$$s = \frac{Q}{2\pi T} \ln c \frac{r_2}{r_1} \tag{6.13-1}$$

where c must now be determined.

To determine c we can use the condition that $s = 0$ at the edge of the aquifer. Thus, along the edge

$$\ln c \frac{r_2}{r_1} = 0$$

which implies that

$$cr_2 = r_1 \tag{6.13-2}$$

Equation (6.13-2) can be written in cartesian coordinates in the form

$$\left(x - \frac{\delta - c^2 d}{1 - c^2} \right)^2 + y^2 = \frac{c^2(d - \delta)^2}{(1 - c^2)^2} \tag{6.13-3}$$

The last equation represents a family of circles along which the drawdown must be zero. Thus, since $s = 0$ for a circle centered at $(0, 0)$ and of radius r_0 it follows that

$$r_0 = \frac{c(d - \delta)}{1 - c^2} \qquad (6.13\text{-}4)$$

and

$$\frac{\delta - c^2 d}{1 - c^2} = 0 \qquad (6.13\text{-}5)$$

Simultaneously solving equations (6.13-4) and (6.13-5) one obtains that

$$c = \delta/r \qquad (6.13\text{-}6)$$

and

$$d = r_0^2/\delta \qquad (6.13\text{-}7)$$

Thus, equation (6.13-1) can now be written in cartesian coordinates and using equations (6.13-6) and (6.13-7) one obtains that

$$s = \frac{Q}{4\pi T} \ln \frac{\delta^2\left[\left(x - \frac{r_0^2}{\delta}\right)^2 + y^2\right]}{r_0^2[(x - \delta)^2 + y^2]} \qquad (6.13\text{-}8)$$

It is suggested as an exercise, that the reader prove that as $\delta \to 0$ equation (6.13-8) reduces to the concentric case, that is,

$$s = \frac{Q}{2\pi T} \ln \frac{r_0}{r} \qquad (6.13\text{-}9)$$

Jacob[8] was interested in comparing the drawdown at the face of an eccentric well with that at the face of a concentric well for equal discharges. Using equations (6.13-8) and (6.13-9) he was able to conclude that

$$\frac{s_{w(\text{concentric})}}{s_{w(\text{eccentric})}} = \frac{\ln \dfrac{r_0}{r_w}}{\ln \dfrac{r_0}{r_w} + \ln \left(1 - \dfrac{x_0^2}{r_0^2}\right)} \qquad (6.13\text{-}10)$$

Using equation (6.13-10) the graph shown on Figure 6.15 can be constructed. This graph shows the drawdown comparison for several values of r_0/r_w.

---

[8] Jacob C. E. (1950) *Flow of Groundwater*, in Engineering Hydraulics, edited by Hunter Rouse, John Wiley & Sons Inc., N.Y.

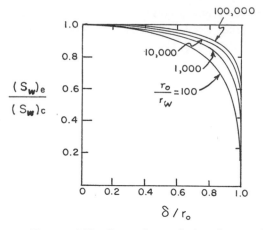

Figure 6.15 Comparison of drawdowns of eccentric and concentric wells of equal discharge (after C. E. Jacob).

Polubarinova-Kochina,[9] on the other hand, compares the discharge for an eccentric well with that from a concentric well for equal drawdowns. Mme. Polubarinova's results can easily be obtained using the above discussion.

6.14 Utilization of the Method of Images in Complex Cases

There are numerous problems not considered here which can be solved using the method of images. For example, this method can be used to solve problems involving multiple wells in a given field, problems involving boundaries such as two rivers intersecting at an angle or a river intersecting an impermeable lens.

Some problems, when solved by the method of images, involve only one real well and a single image mirrored only about one plane. However in many cases an infinitude of images is obtained (see Muskat[1]). In all cases, however, the techniques involved in solving the problems are similar to those described in Sections 6.12 and 6.13. For further details the reader is referred to the work of Muskat,[1] Jacob,[8] Polubarinova,[9] and DeWiest.[10]

[9] Polubarinova-Kochina, P. Ya (1962) *Theory of Groundwater Movement*, Princeton University Press, Princeton, New Jersey.

[10] DeWiest, Roger, J. M. (1965) *Geohydrology*, John Wiley & Sons, Inc., N.Y.

6.15 Influence of Precipitation in the Groundwater Flow Pattern

In the preceding sections the flow into a well has been treated assuming that no recharge into the aquifer directly takes place. In certain cases of unconfined flow this assumption does not hold and one must therefore consider the effect that infiltration and evaporation have on the groundwater flow regime.

Let us consider now the case of an unconfined aquifer between two rivers that is being recharged in its entire surface by an amount of water per unit area W. Then (see Figure 6.16),

$$q = -Kh \frac{\partial h}{\partial x} \tag{6.15-1}$$

and

$$q = \int W \, dx \tag{6.15-2}$$

Consequently,

$$K \frac{d}{dx} \left(h \frac{dh}{dx} \right) = -W \tag{6.15-3}$$

Solving equation (6.15-3) one obtains

$$Kh^2 + Wx^2 = c_1 x + c_2 \tag{6.15-4}$$

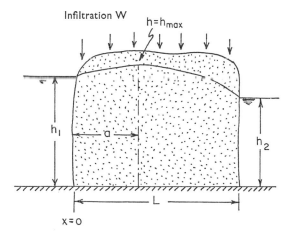

Figure 6.16 An unconfined aquifer between two rivers undergoing direct recharge (infiltration).

This last equation is the representation of an ellipse if $W > 0$ which is the case for infiltration, and if $W < 0$ the equation represents a hyperbola which is the case for evaporation.

For the problem under consideration as shown in Figure 6.16 the boundary conditions are

$$h(0) = h_1 \quad \text{and} \quad h(L) = h_2$$

Using these boundary conditions the constants c_1 and c_2 in the preceding equation can be evaluated to obtain

$$h = \left(h_1{}^2 - \frac{(h_1{}^2 - h_2{}^2)x}{L} + \frac{W}{K}(L - x)x \right)^{1/2} \tag{6.15-5}$$

The point where h is a maximum is called the *groundwater divide*. All the water that filters to the left of the divide flows to the left while that water which filters to the right flows to the right. The distance a between $x = 0$ and the divide can be obtained using the formula

$$a = \frac{L}{2} - \frac{K}{W}\frac{h_1{}^2 - h_2}{2L} \tag{6.15-6}$$

Thus, for an aquifer located between two rivers undergoing uniform infiltration equation (6.15-5) gives us the head profile and equation (6.15-6) gives us the position of the groundwater divide. The divide will be located nearest to the river with the highest water level.

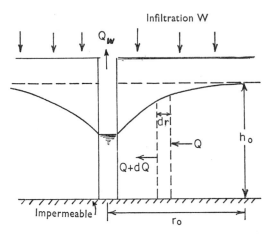

Figure 6.17 Steady flow into a well fully penetrating an unconfined aquifer and including direct infiltration.

A very important case is that of a well in an unconfined aquifer in an area of heavy rains where direct infiltration makes up a considerable amount of the water fed to the aquifer. In this case (see Figure 6.17)

$$dQ = -2W\pi r\, dr \qquad (6.15\text{-}7)$$

and thus,

$$h_0{}^2 - h^2 = \frac{W}{2K}(r^2 - r_0{}^2) + \frac{Q}{\pi K}\ln\frac{r_0}{r} \qquad (6.15\text{-}8)$$

The term $(W/2K)(r^2 - r_0{}^2)$ in equation (6.15-8) represents the contribution made by infiltration.

6.16 The Complex Variable Method

The theory of functions of a complex variable is extremely useful for solving hydrologic problems involving the potential functions ϕ and ψ. In this section we will illustrate the use of complex variables to solve the problem of flow into and out of two wells one recharging and the other discharging water from a given aquifer. Before reading this section it will be useful that the reader familiarize himself with the language of complex variables.[11,12,13,14]

We begin by recalling the definition of the complex potential as given in Section 4.3, that is,

$$\omega(z) = \phi(z) + i\psi(z) \qquad (6.16\text{-}1)$$

where $z = x + iy$ and $i = \sqrt{-1}$.

Consider now two wells, one recharging and the other discharging the aquifer as shown on Figure 6.18. The recharge Q of the one is equal to the other's discharge and they are separated a distance $2x_1$. For this case the complex potential ω at a point z is

$$\omega(z) = -\frac{q}{2\pi T}\ln\left(\frac{z + x_1}{z - x_1}\right) \qquad (6.16\text{-}2)$$

To simplify matters we will let $T = 1$ and transform equation (6.16-2)

[11] Whittaker, E. T. and Watson, G. N. (1927) *A Course in Modern Analysis*, Cambridge University Press, New York USA.

[12] McLachlan, N. W. (1939) *Complex Variable and Operational Calculus*, Cambridge University Press, New York, USA.

[13] Sokolnikoff, I. S. and Redheffer, R. M. (1958) *Mathematics of Physics and Modern Engineering*, McGraw-Hill Book Co. New York, USA.

[14] MacRobert, T. M. (1933) *Functions of a Complex Variable*, The Macmillan Company, New York, USA.

to polar coordinates using the transformation

$$z = r \exp(i\theta) = r(\cos\theta + i\sin\theta) \tag{6.16-3}$$

We then obtain that

$$\omega(z) = -\frac{q}{2\pi}\ln\frac{r'}{r} + i\left(-\frac{q}{2}(\theta' - \theta)\right) \tag{6.16-4}$$

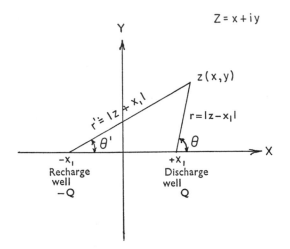

Figure 6.18   Two-well system.

Therefore,

$$\phi = -\frac{q}{2\pi}\ln\frac{r'}{r} = \frac{q}{4\pi}\ln\frac{(x-x_1)^2 + y^2}{(x+x_1)^2 + y^2} \tag{6.16-5}$$

$$\psi = -\frac{q}{2\pi}(\theta' - \theta) = \frac{q}{2\pi}\left(\tan^{-1}\frac{y}{x - x_1} - \tan^{-1}\frac{y}{x + x_1}\right) \tag{6.16-6}$$

Equations (6.16-5) and (6.16-6) allow the construction of a flow net representative of the case under consideration. Using equation (6.16-5) we can see that equipotential lines must satisfy the condition $\phi = $ constant which for this case implies that

$$e^{4\pi\phi/q} = \frac{(x-x_1)^2 + y^2}{(x+x_1)^2 + y^2} = a \tag{6.16-7}$$

Consequently,

$$x^2 - 2x_1x\left(\frac{1+a}{1-a}\right) + x_1^2\left(\frac{1+a}{1-a}\right)^2 + y^2 = x_1\left(\left(\frac{1+a}{1-a}\right)^2 - 1\right) \tag{6.16-8}$$

As shown on Figure 6.19 equation (6.16-8) represents a family of circles with radii

$$r = 2x_1\sqrt{a}/(1 - a)$$

and centers in the points that satisfy the conditions

$$x = x_1\left(\frac{1 + a}{1 - a}\right), \qquad y = 0$$

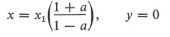

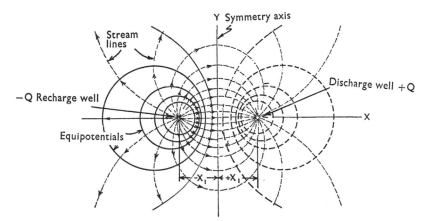

Figure 6.19   Flow net for the case of one recharge and one discharge well separated a distance $2x_1$.

To obtain the streamlines one can use equation (6.16-6) and apply the condition $\psi = $ constant. Consequently,

$$\tan\frac{2\pi\psi}{q} = \tan(\theta - \theta') = \frac{\dfrac{y}{x - x_1} - \dfrac{y}{x + x_1}}{1 + \left(\dfrac{y}{x - x_1}\right)\left(\dfrac{y}{x + x_1}\right)} = b$$

which reduces to

$$x^2 + \left(y - \frac{x_1}{b}\right)^2 = x_1^2 + \frac{x_1^2}{b^2} \qquad (6.16\text{-}9)$$

This equation represents a family of circles with centers at the points

$$x = 0, \qquad y = x_1/b$$

and radii

$$r = x_1\sqrt{1 + (1/b^2)}$$

Equations (6.16-8) and (6.16-9) fully represent the flow net for the case under consideration. Using these equations one obtains Figure 6.19. This figure permits us to visualize the flow regime for the system in question and to evaluate the interference between two wells.

6.17 Multiple Wells

An important case not yet considered is the extraction of oil or groundwater from a single aquifer where a collection of n wells each producing an amount Q_i fully penetrate the aquifer. Morris Muskat[1] originally developed, rather extensively, the theory of multiple well fields. Muskat's theory has been applied mostly to the use of multiple wells in oil exploitation and has been extended rather recently by the work of Matthews and Russell.[15]

When n wells fully penetrate a confined aquifer the drawdown at a point P can be obtained by applying the superposition theorem to Thiem's equation and therefore,

$$s = h_0 - h = \sum_{i=1}^{n} \frac{Q_i}{2\pi T} \ln \frac{R_i}{r_i} \qquad (6.17\text{-}1)$$

where R_i is the distance from the ith well to the point where the drawdown is zero and r_i is the distance from the ith well to the point P.

Similarly, one can see that for n wells fully penetrating an unconfined aquifer

$$h_0^2 - h^2 = \sum_{i=1}^{n} \frac{Q_i}{\pi K} \ln \frac{R_i}{r_i} \qquad (6.17\text{-}2)$$

provided the total drawdown is small.

Muskat, and Matthews and Russell have also calculated the interference between several wells arranged according to some geometrical pattern. For example, it can be shown that two wells separated a distance d mutually interfere in such a way that the discharge from each well is

$$Q = Q_1 = Q_2 = \frac{2\pi T(h_0 - h_w)}{\ln (r_0^2/r_w d)} \qquad r_0 \gg d \qquad (6.17\text{-}3)$$

Three wells lined up a distance d apart as in Figure 6.20 interfere mutually so that the middle well discharges

$$Q_B = \frac{2\pi T(h_0 - h_w) \ln (d/2r_w)}{2 \ln (r_0/d) \ln (d/r_w) + \ln (d/2r_w) \ln (r_0/r_w)} \qquad (6.17\text{-}4)$$

---

[15] Matthews, C. S. and Russell, D. G. (1967) *Pressure Buildup and Flow Tests in Wells*, Monograph vol. 1, Henry L. Doherty Memorial Fund, AIME, N.Y.

and the two outer wells produce a discharge

$$Q_A = Q_C = \frac{2\pi T(h_0 - h_w) \ln (d/r_w)}{2 \ln (r_0/d) \ln (d/r_w) + \ln (d/2r_w) \ln (r_0/r_w)}$$ (6.17-5)

To illustrate how one can calculate the extent of interference between two or more wells let us consider a simple example. Suppose that there are two wells in a given field (see Figure 6.21) located at points (x_1, y_1) and (x_2, y_2) respectively and one desires to calculate the interference between them.

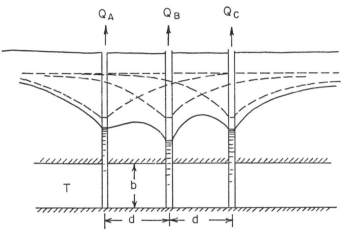

Figure 6.20   A system of three adjacent wells.

The first step is to transfer the wells so that $y_1 = y_2 = 0$ as done in Figure 6.21.

For purposes of illustration let us assume that the aquifer penetrated by the wells is fully confined and the wells fully penetrate the aquifer. Then, if it is known that for the aquifer in question the wells are 150 feet apart and

$$K = 10^{-1} \text{ ft/min}$$
$$b = 60 \text{ ft}$$
$$Q = 60 \text{ ft}^3/\text{min}$$
$$r_0 = 1500 \text{ ft}$$
$$r_w = 1 \text{ ft}$$
$$h_0 = 70 \text{ ft}$$

then one can use formula (6.17-1) to determine the head at a given point in the zone of influence of the two wells.

Using cartesian coordinates and applying formula (6.17-1) for the two wells above one obtains that

$$h(x, 0) = \frac{Q}{2\pi T} \ln (x + d/2)(x - d/2) + C$$

Thus,

$$h(x, 0) = 1.59 \ln (x + 75)(x - 75) + C$$

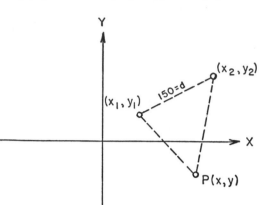

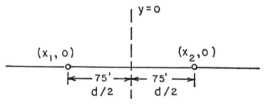

Figure 6.21   Two wells separated a distance d.

The quantity C can be determined applying the conditions existing at r_0. Then, one obtains that $C = 46.6$ ft. Consequently, the drawdown at the well face of either well is

$$s_w = s(76', 0) = 15.4 \text{ ft}$$

If instead of the system of two wells considered above each of them were alone, the drawdown at the well face would be

$$s_w = \frac{Q}{2\pi T} \ln \frac{r_0}{r_w} = 11.7 \text{ ft}$$

Consequently, the drawdown in either of the two wells increases 3.7 ft (roughly 30%) due to the presence of the other well.

It is economically very important to calculate the degree of well interference when considering the possibility of using multiple wells to exploit a given oil or water reservoir. As illustrated in the example above the degree of interference between two adjacent wells can be rather large and can reach the point of economic inefficiency unless the various wells are properly positioned.

6.18 Tridimensional Flow

All cases of flow considered so far have involved one or two dimensions. There are certain instances when hydrologic problems must be solved in three dimensions and no further approximations are possible. For example, one such case is the problem of flow into a well of small radius just tapping a thick confined aquifer. In this case there is spherical symmetry with respect to an axis passing through the center of the well. Thus, under steady state conditions and given the symmetry of the problem, Laplace's tridimensional equation in spherical coordinates can be simplified to

$$\frac{1}{r^2}\frac{\partial}{\partial r}\left(r^2\frac{\partial h}{\partial r}\right) = 0 \qquad (6.18\text{-}1)$$

The solution to equation (6.18-1) is

$$h = -\frac{C_1}{r} + C_2 \qquad (6.18\text{-}2)$$

As before we can use the boundary conditions $h = h_w$ at $r = r_w$ and $h = h_0$ at $r = r_0$. Applying these conditions to equation (6.18-2) we obtain that

$$h = h_w + \frac{h_0 - h_w}{(1/r_0) - (1/r_w)}\left(\frac{1}{r} - \frac{1}{r_w}\right) \qquad (6.18\text{-}3)$$

In order to find the fluid velocity v_r we can use Darcy's law. Thus,

$$v_r = -K\frac{\partial h}{\partial r} = K\frac{h_0 - h_w}{(1/r_0) - (1/r_w)}\frac{1}{r^2} \qquad (6.18\text{-}4)$$

To calculate Q we use equation (6.18-4) and obtain

$$Q = -\int_0^{2\pi} d\chi \int_0^{\pi} r^2 \sin\theta v_r \, d\theta = 4\pi K\frac{h_0 - h_w}{(1/r_w) - (1/r_0)} \qquad (6.18\text{-}5)$$

Consequently, combining equations (6.18-3) and (6.18-5) we obtain that

$$h = \frac{Q}{4\pi K}\left(\frac{1}{r_w} - \frac{1}{r}\right) + h_w \qquad (6.18\text{-}6)$$

and since $r_w \ll r_0$ it follows that equation (6.18-5) can be reduced to

$$Q \approx 4\pi K(h_0 - h_w)r_w \qquad (6.18\text{-}7)$$

This last formula indicates that if a well just taps an aquifer the discharge obtained is independent of the radius of influence.

6.19   Partial Penetration

The preceding section discussed flow into a well partially penetrating a given aquifer. In certain instances partial penetration of an aquifer is desirable. For example, if one is extracting water from an aquifer containing salt water at the bottom it is convenient only to penetrate it a distance h_s such that the salt water cone produced upon pumping will remain stable and will not contaminate the fresh water being discharged.

Muskat[1] and de Glee[16] have considered the problem of partial penetration in great detail. For a well partially penetrating a confined aquifer (see Figure

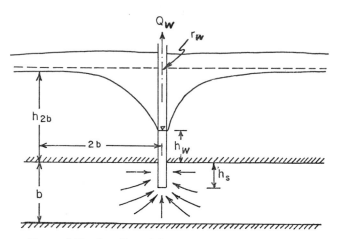

Figure 6.22   A well partially penetrating a confined aquifer.

---

[16] de Glee, G. J. (1930) *Over Grondwaterstromingen by Wateronttrekking by Middel van Putten*, T. Waltman J. Delft.

6.22) de Glee has shown that

$$h_{2b} - h_w = \frac{Q_w}{4\pi K}\left(\frac{2}{h_s}\ln\frac{\pi h_s}{2r_w} + \frac{0.20}{b}\right) \tag{6.19-1}$$

for $1.3h_s \leq b$ and $h_s/2r_w \geq 5$.

The above formula allows us to find h_{2b}. In order to find the total drawdown $h - h_w$ we must add to equation (6.19-1) the drawdown beyond $2b$.

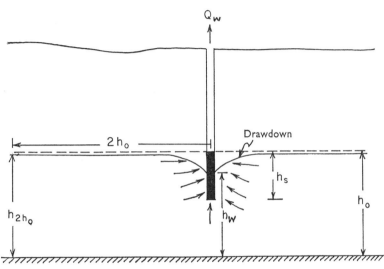

Figure 6.23 A well partially penetrating an unconfined aquifer.

This can be accomplished on the assumption that beyond $2b$ it does not matter whether the well is partially or fully penetrating. Thus, one obtains that

$$h - h_w = \frac{Q_w}{4\pi K}\left(\frac{2}{h_s}\ln\frac{\pi h_s}{2r_w} + \frac{0.20}{b} + \frac{2}{b}\ln\frac{r}{2b}\right)$$

Similarly, for the case of a well partially penetrating an unconfined aquifer one obtains that

$$h_{2h_0} - h_w = \frac{Q_w}{4\pi K}\left(\frac{2}{h_s}\ln\frac{\pi h_s}{2r_w} + \frac{0.20}{h_0}\right) \tag{6.19-2}$$

The formula (6.19-2) is similar to (6.19-1) the only difference being the use of h_0 instead of b. The case of partial penetration of an unconfined aquifer is depicted in Figure 6.23.

6.20 Principle of Superposition

In 1937 Wenzel[17] considered an important hydrologic problem. He was interested in finding the permeability of an unconfined aquifer where the water table is inclined an angle α with respect to the horizontal. Figure 6.24 shows such a situation. The angular displacement has been exaggerated in

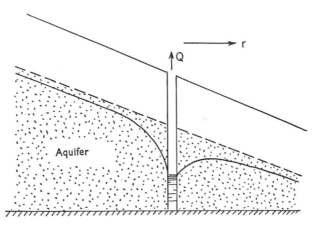

Figure 6.24 An unconfined aquifer with a sloping water table.

this figure to show its effect on the head profile more clearly. Wenzel found that for this case

$$K = \frac{2Q}{r(h_u + h_d)(i_u + i_d)} \qquad (6.20\text{-}1)$$

where r = distance from the pumping well,

h_u, h_d = saturated thicknesses up and down hill a distance r from the well, and

i_u, i_d = slopes of the initial water table up and down hill at a distance r from the well.

Now let us consider the case of a confined aquifer with a sloping piezometric surface (see Figure 6.25) and show how the flow net for this case can be constructed and how the zone contributing water into the well can be delimited.

---

[17] Wenzel, L. K. (1937) *The Thiem Method for Determining Permeability of Water Bearing Materials*, U.S.G.S. Survey Paper #679.

For unidirectional flow it is obvious that $\phi_u = v_u x$ and $\psi_u = v_u y$. Also, we have seen that for flow into a discharging well

$$\phi_w = \frac{q}{2\pi} \ln \frac{r}{r_w}$$

$$\psi_w = \frac{q\theta}{2\pi}$$

Consequently, applying the superposition principle we obtain that

$$\phi = v_u x + \frac{q}{2\pi} \ln \frac{r}{r_w} \qquad\qquad (6.20\text{-}2)$$

$$\psi = v_u y + \frac{q\theta}{2\pi} \qquad\qquad (6.20\text{-}3)$$

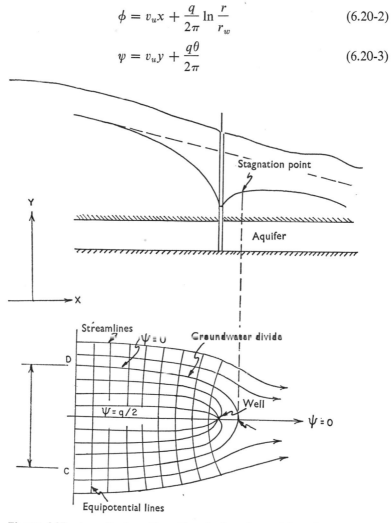

Figure 6.25 A confined aquifer with a sloping piezometric surface.

Using the last two equations we can construct a flow net for the situation as shown on Figure 6.25. Point A in Figure 6.25 is called the *stagnation point*. At this point $|v_u| = |v_w|$. Therefore at the stagnation point

$$\frac{q}{2\pi x_A} = K\frac{dh}{dx}$$

and therefore the separation between the well and the stagnation point, x_A, is

$$x_A = \frac{q}{2\pi K(dh/dx)} \tag{6.20-4}$$

Equation (6.20-4) permits us to determine the exact position of the stagnation point.

The zone contributing water into the well is limited by the groundwater divide as shown on Figure 6.25. To completely delimit this zone we need to know not only the position of the stagnation point but also the distance CD. To find this distance we note that at the point D, $\psi = 0$ and $\theta = \pm\pi$. Consequently $2y = \pm q/v_u$ where $CD = |2y|$.

The flow net shown in Figure 6.25 not only defines clearly the zone contributing water into the well but also gives us a picture of the groundwater flow regime in the well's vicinity.

Exercises

1. Consider a 2000 meter long confined aquifer located between two rivers. If the water level in one of the rivers is 50 meters above the impermeable base while the water level on the other is 87 meters above the base, then, construct a graph of the piezometric surface in the aquifer as a function of the space coordinate x.

2. Repeat problem 1 for the case of an unconfined aquifer.

3. Repeat problem 1 for the case of a leaky aquifer. Assume that $h_0 = 60$ meters, and $B = 10^4$ meters.

4. Beginning with

$$\frac{\partial^2 h}{\partial x^2} + \frac{h_0 - h}{B^2} = 0$$

derive that $h = h_0 + c_1 \cosh x/B + c_2 \sinh x/B$.

5. Consider a confined aquifer of variable thickness (see Figure 6.5). Given that $h_1 = 40$ m, $h_2 = 16$ m, $b_0 = 12$ m, $C = 1$ m^{-1}, and $l = 500$ m, construct a graph of h versus x.

6. Calculate q in Problem 5, if the transmissivity is 100 meters^2/day.

7. Discuss in detail the physical significance of Dupuit's conditions.

8. Solve the equation

$$h \frac{\partial^2 h^2}{\partial x^2} - m \frac{\partial h^2}{\partial x} = 0$$

9. Consider an unconfined aquifer 12 meters thick. A water well fully penetrates the aquifer producing during a 72-hour pumping test a discharge of 18 liters/second. Near the well, two observation wells are located 10 meters and 25 meters away from the pumping well. During the pumping test considered above drawdowns of 1.50 m and 0.50 m were observed in the two observation wells respectively. Calculate the aquifer's permeability assuming that $h_0 = 12$ meters.

10. Repeat the preceding problem for the case of a confined aquifer.

11. What is the meaning of the term specific capacity and how does it relate to the efficiency of a well?

12. What is the significance of Figure 6.15?

13. For the case of uniform infiltration into an aquifer located between two rivers show that the groundwater divide is nearest to the river whose water level is highest.

14. Determine W/K and h_{max} for the unconfined aquifer between two rivers shown in the figure below. Assume the direct infiltration is uniform.

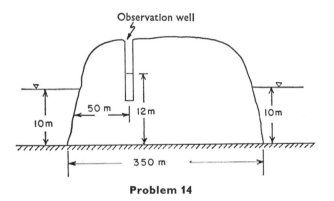

Problem 14

15. Consider two wells fifty meters apart, one producing a discharge Q and the other recharging the aquifer an amount Q. Construct a scaled flow net to represent this situation.

16. Compare the solution to the flow equation for a well just tapping a confined aquifer with that of a fully penetrating well. What are the most striking differences and similarities?

17. When is it preferable to only penetrate partially an aquifer?

18. Consider a well drilled in an aquifer where there is unidirectional flow in the positive x direction. Given that

$$\psi = K \frac{dh}{dx} y - \frac{q\theta}{2\pi} = v_x y - \frac{q\theta}{2\pi}$$

(a) Calculate the value of the equipotential line that intersects the stagnation point.

(b) Determine the value of the streamlines along $y = 0$.

19. Consider a well drilled in an aquifer where there is unidirectional flow in the negative x direction. For this situation

(a) determine the function ψ along the groundwater divide, and

(b) determine the position of the stagnation point.

CHAPTER 7

Unsteady Flow

In the preceding chapter many hydrologic problems concerning the flow of groundwater into a well under equilibrium conditions were studied in detail. There are also numerous field situations where time must be considered because an equilibrium condition has not been fully achieved. Such problems are grouped here under the heading of "Unsteady Flow."

C. V. Theis[1] pioneered this field by obtaining the solution to the problem of unsteady flow from a confined aquifer into a well. He solved this and a few other problems using as an analogy the flow of heat. Jacob continued and further extended Theis' work using well-known hydrologic principles. Finally, Hantush extended Theis' theories to leaky aquifers.

This chapter will be concerned with the fundamental theories of unsteady flow through a porous medium and into a well. Also, various graphical and mathematical techniques that can be used to determine the formation constants (storage and permeability) of an aquifer under unsteady conditions will be analyzed.

7.1 Flow into a Well Penetrating a Uniformly Thick Confined Aquifer

Consider now the case of a well fully-penetrating a confined, horizontal, uniformly-thick aquifer. For the case of unsteady flow one must apply the differential equation

$$\frac{\partial^2 h}{\partial x^2} + \frac{\partial^2 h}{\partial y^2} + \frac{\partial^2 h}{\partial z^2} = \frac{S}{T}\frac{\partial h}{\partial t} \tag{7.1-1}$$

where h is the hydraulic head.

---

[1] Theis, C. V. (1935) *The Relation Between the Lowering of the Piezometric Surface and the Rate and Duration of Discharge of a Well Using Groundwater Storage*, Trans. A.G.U., vol. 16, pp. 519–524.

Equation (7.1-1) can be written in terms of the drawdown, s ($s = h_0 - h$), as

$$\frac{\partial^2 s}{\partial x^2} + \frac{\partial^2 s}{\partial y^2} + \frac{\partial^2 s}{\partial z^2} = \frac{S}{T}\frac{\partial s}{\partial t} \tag{7.1-2}$$

For the case considered here (see Figure 7.1) it is more practical for reasons of symmetry to transform equation (7.1-2) into cylindrical coordinates.

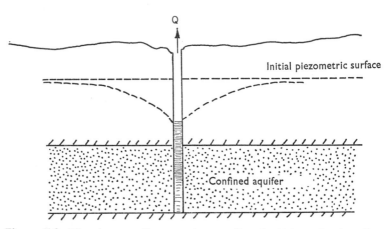

Figure 7.1 Flow into a well penetrating a uniformly-thick confined aquifer.

Also, because of symmetry one can disregard drawdown variations with respect to z and θ (see Section 3.20). Thus, equation (7.1-2) becomes

$$\frac{\partial^2 s}{\partial r^2} + \frac{1}{r}\frac{\partial s}{\partial r} = \frac{1}{v}\frac{\partial s}{\partial t} \tag{7.1-3}$$

where $v = T/S$.

Therefore, the problem consists in solving equation (7.1-3) subject to conditions

$$s(\infty, t) = 0 \qquad t > 0 \tag{7.1-4}$$

$$s(r, 0) = 0, \tag{7.1-5}$$

and

$$\lim_{r \to 0} r\frac{\partial s}{\partial r} = -\frac{Q}{2\pi T} \tag{7.1-6}$$

Equation (7.1-3) can be solved by the method of separation of variables, that is, letting

$$s = R(r)T(t)$$

By this method one obtains that

$$s = \frac{Q}{4\pi T} \int_{r^2S/4Tt}^{\infty} \frac{e^{-x}}{x} \, dx$$

$$s = \frac{Q}{4\pi T} W\left(\frac{r^2S}{4Tt}\right) = \frac{QW(u)}{4\pi T} \tag{7.1-7}$$

where $u = r^2S/4Tt$.

The function $W(u)$ is known as the *Well Function*. It can be expanded in series form as

$$W(u) = -0.5772 - \ln u + u - \frac{u^2}{2!\,2} + \frac{u^3}{3!\,3} - \frac{u^4}{4!\,4} + \cdots \tag{7.1-8}$$

Table 7.1 contains values of $W(u)$ as a function of u. Figure 7.2 presents the Well Function in graphical form.

After a given time has elapsed some terms in the series expansion of the Well Function become very small and do not affect the value of the drawdown. In such cases the Well Function can be approximated by the equation

$$W(u) = \ln \frac{2.25Tt}{Sr^2} \qquad u < 0.02 \tag{7.1-9}$$

Consequently, if $u < 0.02$

$$s = \frac{Q}{4\pi T} \ln \frac{2.25Tt}{Sr^2} \tag{7.1-10}$$

If one desires to calculate the drawdown in a given point in the vicinity of a well after a given pumping time, then the formula (7.1-10) can be modified and written in the form

$$s_2 - s_1 = \frac{2.3Q}{4\pi T} \log \frac{r_1^2/t_1}{r_2^2/t_2} \tag{7.1-11}$$

where s_2 and s_1 are the drawdowns at points located distances r_2 and r_1 from the well after times t_2 and t_1 have elapsed from the beginning of pumping.

7.2 Theis' Method of Calculating Formation Constants

In the previous section the drawdown due to pumping water from a confined aquifer under unsteady state conditions has been determined. In his original analysis of this problem Theis' purpose was to determine the storage coefficient and the transmissivity of confined aquifers. The method he devised to

REGIONAL HYDROLOGY FUNDAMENTALS

Table 7.1 The Well Function (after Wenzel, 1942).

N	×1	×10⁻¹	×10⁻²	×10⁻³	×10⁻⁴	×10⁻⁵	×10⁻⁶	×10⁻⁷
1.0	0.22	1.82	4.04	6.33	8.63	10.94	13.24	15.54
1.5	0.10	1.46	3.64	5.93	8.23	10.53	12.83	15.14
2.0	0.05	1.22	3.35	5.64	7.94	10.24	12.55	14.85
2.5	0.03	1.04	3.14	5.42	7.72	10.02	12.32	14.62
3.0	0.013	0.91	2.96	5.23	7.53	9.84	12.14	14.44
3.5	0.0070	0.79	2.81	5.08	7.38	9.68	11.99	14.29
4.0	0.0038	0.70	2.68	4.95	7.25	9.55	11.85	14.15
4.5	0.0021	0.63	2.57	4.83	7.13	9.43	11.73	14.04
5.0	0.0011	0.56	2.47	4.73	7.02	9.33	11.63	13.93
5.5	0.00064	0.50	2.38	4.63	6.93	9.23	11.53	13.84
6.0	0.00036	0.45	2.30	4.54	6.84	9.14	11.45	13.75
6.5	0.00020	0.41	2.22	4.47	6.76	9.06	11.37	13.67
7.0	0.00012	0.37	2.15	4.39	6.69	8.99	11.29	13.60
7.5	0.000066	0.34	2.09	4.32	6.62	8.92	11.22	13.53
8.0	0.000038	0.31	2.03	4.26	6.55	8.86	11.16	13.46
8.5	0.000022	0.28	1.97	4.20	6.49	8.80	11.10	13.40
9.0	0.000012	0.26	1.92	4.14	6.44	8.74	11.04	13.34

N	×10⁻⁸	×10⁻⁹	×10⁻¹⁰	×10⁻¹¹	×10⁻¹²	×10⁻¹³	×10⁻¹⁴	×10⁻¹⁵
1.0	17.84	20.15	22.45	24.75	27.05	29.36	31.66	33.96
1.5	17.44	19.74	22.04	24.35	26.65	28.95	31.25	33.56
2.0	17.15	19.45	21.76	24.06	26.36	28.66	30.97	33.27
2.5	16.93	19.23	21.53	23.83	26.14	28.44	30.74	33.05
3.0	16.74	19.05	21.35	23.65	25.96	28.26	30.56	32.86
3.5	16.59	18.89	21.20	23.50	25.80	28.10	30.41	32.71
4.0	16.46	18.76	21.06	23.36	25.67	27.97	30.27	32.58
4.5	16.34	18.64	20.94	23.25	25.55	27.85	30.15	32.46
5.0	16.23	18.54	20.84	23.14	25.44	27.75	30.05	32.35
5.5	16.14	18.44	20.74	23.05	25.35	27.65	29.95	32.26
6.0	16.05	18.35	20.66	22.96	25.26	27.56	29.87	32.17
6.5	15.97	18.27	20.58	22.88	25.18	27.48	29.79	32.09
7.0	15.90	18.20	20.50	22.81	25.11	27.41	29.71	32.02
7.5	15.83	18.13	20.43	22.74	25.04	27.34	29.64	31.95
8.0	15.76	18.07	20.37	22.67	24.97	27.28	29.58	31.88
8.5	15.70	18.01	20.31	22.61	24.91	27.22	29.52	31.82
9.0	15.65	17.95	20.25	22.55	24.86	27.16	29.46	31.76

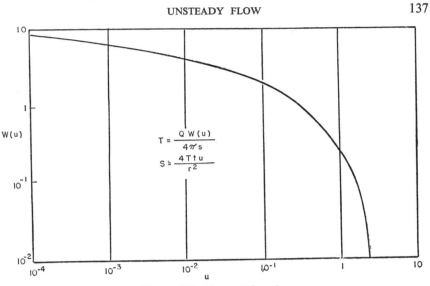

Figure 7.2 The well function.

determine S and T is relatively simple. To calculate the transmissivity T, Theis used formula (7.1-7) directly, that is,

$$T = \frac{QW(u)}{4\pi s} \qquad (7.2\text{-}1)$$

To calculate S, the definition of u can be used, that is,

$$S = \frac{4Ttu}{r^2} \qquad (7.2\text{-}2)$$

In order to visualize this method more clearly consider now a large, uniformly-thick aquifer. Fully penetrating this aquifer is a well producing 100 m³/hour (0.028 m³/sec) of water. From an observation well located 50 meters away from the producing well the drawdown is monitored. During such a pumping test we observe the drawdown s at the observation well resulting from pumping an amount Q at the producing well. During the pumping test one will obtain the following data:

time (min)	drawdown (meters)	meters²/minute
t	s	r^2/t
.	.	.
.	.	.
.	.	.

Using this data a graph of s versus r^2/t is constructed using log-log paper. Figure 7.3 shows such a graph for the pumping test being considered here.

Now, Figure 7.3 is superimposed on Figure 7.2 (graph of $W(u)$ versus u) and the point or points of contact of both curves are noted. The super-position process is shown on Figure 7.4. In this fashion one obtains for Figure 7.4 that

$$W(u) = 2.20 \times 10^{-1}$$

$$u = 1$$

$$s = 2.20 \times 10^{-2} \text{ meters}$$

$$r^2/t = 360 \text{ m}^2/\text{minute} = 6 \text{ m}^2/\text{sec}$$

and consequently using (7.2-1) and (7.2-2) we obtain that

$$T = 2.2 \times 10^{-2} \text{ m}^2/\text{sec}$$

and

$$S = 0.0147$$

7.3 Jacob's Method of Calculating Formation Constants

Jacob[2] devised a method to determine S and T using the approximation

$$s = \frac{Q}{4\pi T} \ln \frac{2.25Tt}{r^2 S} = \frac{2.30Q}{4\pi T} \log \frac{2.25Tt}{r^2 S} \qquad (7.3\text{-}1)$$

To illustrate this technique the same pumping test considered in the previous section will now be used. First, one must plot s versus t/r^2 in semi-logarithmic paper (see Figure 7.5). Thus,

$$\frac{t_0}{r^2} = 5 \times 10^{-3} \text{ min/m}^2 = 3 \times 10^{-1} \text{ sec/m}^2$$

and

$$\Delta h = 0.23 \text{ meters}$$

From equation (7.3-1) and given t_0 and Δh one can use the formulas

$$T = \frac{0.183Q}{\Delta h}$$

and

$$S = \frac{2.25Tt_0}{r^2}$$

---

[2] Jacob, C. E. (1950) *Flow of Groundwater*, Ch. 5 of Engineering Hydraulics, Hunter Rouse ed., John Wiley & Sons, N.Y.

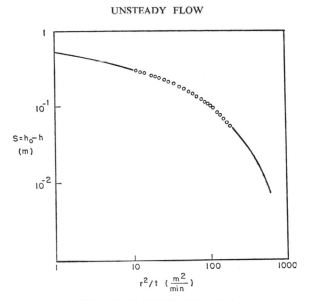

Figure 7.3 The Theis' method.

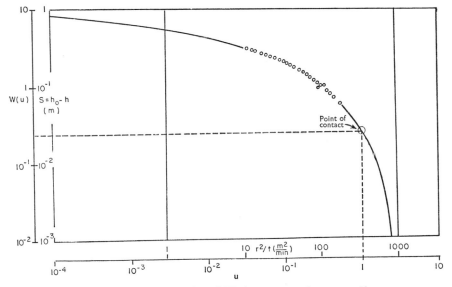

Figure 7.4 Superposition of $W(u)$ versus u and s versus r^2/t.

Consequently, for the case under consideration

$$T = 2.23 \times 10^{-2} \text{ m}^2/\text{sec}$$

$$S = 1.51 \times 10^{-2}$$

These results are in excellent agreement with those obtained in the previous section.

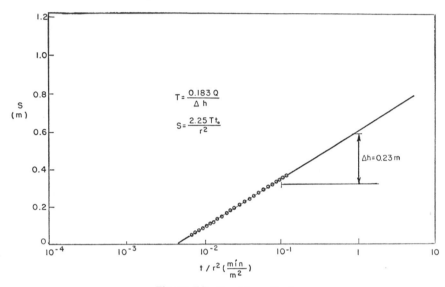

Figure 7.5   Jacob's method.

7.4   Flow Into a Well Penetrating a Leaky Aquifer

The general equation that describes groundwater flow in a leaky aquifer has already been shown to be

$$\frac{\partial^2 s}{\partial r^2} + \frac{1}{r}\frac{\partial s}{\partial r} - \frac{s}{B^2} = \frac{1}{\nu}\frac{\partial s}{\partial t} \tag{7.4-1}$$

where s is the drawdown.

We are interested here in solving the problem of unsteady flow into a well fully penetrating a leaky aquifer such as depicted on Figure 7.6. Such a case can be considered by solving equation (7.4-1) subject to conditions

$$s(r, 0) = 0, \tag{7.4-2}$$

$$s(\infty, t) = 0, \qquad t > 0, \tag{7.4-3}$$

and

$$\lim_{r \to 0} r \frac{\partial s}{\partial r} = -\frac{Q}{2\pi T} \qquad (7.4\text{-}4)$$

Applying the Laplace Transformation to (7.4-1), (7.4-3), and (7.4-4) we obtain

$$\frac{\partial^2 \bar{s}}{\partial r^2} + \frac{1}{r} \frac{\partial \bar{s}}{\partial r} - \left(\frac{1}{B^2} + \frac{p}{v}\right)\bar{s} = 0, \qquad (7.4\text{-}5)$$

$$\bar{s}(\infty, p) = 0, \qquad (7.4\text{-}6)$$

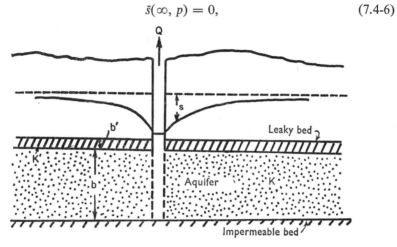

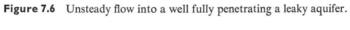

Figure 7.6 Unsteady flow into a well fully penetrating a leaky aquifer.

and

$$\lim_{r \to 0} r \frac{\partial \bar{s}}{\partial r} = -\frac{Q}{2\pi T p} \qquad (7.4\text{-}7)$$

Equation (7.4-5) is a modified form of Bessel's equation and its solution is

$$\bar{s} = c_1 I_0\left(r\sqrt{\frac{1}{B^2} + \frac{p}{v}}\right) + c_2 K_0\left(r\sqrt{\frac{1}{B^2} + \frac{p}{v}}\right) \qquad (7.4\text{-}8)$$

where the constants c_1 and c_2 can be evaluated using (7.4-6) and (7.4-7). Thus,

$$\bar{s} = \frac{Q}{2\pi T p} K_0\left(r\sqrt{\frac{1}{B^2} + \frac{p}{v}}\right) \qquad (7.4\text{-}9)$$

Now, we must find the inverse transform, that is,

$$s = \frac{Q}{2\pi T} \mathscr{L}^{-1}\left\{\frac{1}{p} \cdot K_0\left(r\sqrt{\frac{1}{B} + \frac{p}{v}}\right)\right\} \qquad (7.4\text{-}10)$$

To solve this last expression we must look up in a table of Laplace transforms[3] the following two transformations

$$\mathscr{L}^{-1}\left\{\frac{1}{p}\right\} \quad \text{and} \quad \mathscr{L}^{-1}K_0\left(r\sqrt{\frac{1}{B^2} + \frac{p}{v}}\right) \tag{7.4-11}$$

Then, applying these to equation (7.4-10) and using the convolution theorem one obtains that

$$s = \frac{Q}{4\pi T} W(u, \beta) = \frac{Q}{4\pi T} \int_u^\infty \frac{\exp\left(\xi + \dfrac{r^2/B^2}{4\xi}\right)}{\xi} d\xi \tag{7.4-12}$$

where $u = r^2S/4tT$ and $\beta = r/B$. This last formula is known as *the Hantush formula for leaky aquifers* and the function $W(u, \beta)$ which is tabulated in Table 7.2 is *the Well Function For Leaky Aquifers*. Hantush and Jacob[4] discuss the well function for leaky cases in further detail and give additional information regarding the above derivation.

7.5 Determination of the Formation Constants of a Leaky Aquifer

In the preceding section the formula for the drawdown in a leaky aquifer has been shown to be

$$s = \frac{Q}{4\pi T} W(u, \beta) \tag{7.5-1}$$

The function $W(u, \beta)$ was represented in the previous section as an integral which can in turn be expanded into a rather cumbersome series form and is best represented by the tabular form of Table 7.2. The function can also be graphed in log-log paper as a function of $1/u$ for various values of β. Such a graph is shown on Figure 7.7 and will be used in determining S and T. These curves are called "Leaky Aquifer Type Curves."

To calculate S and T we must graph s versus r^2/t as done previously in Section 7.2. Then, knowing β we select the corresponding type curve from

---

[3] Churchill, R. V. (1958) *Operational Mathematics*, McGraw-Hill Book Company, New York.

[4] Hantush, M. S. and Jacob, C. E. (1955) *Nonsteady radial flow in an infinite leaky aquifer and nonsteady Green's functions for an infinite strip of leaky aquifer*, Trans. AGU, vol. 36, pp. 95–112.

Table 7.2 The function $W(u, \beta)$. Courtesy of New Mexico Institute of Mining and Technology, from M. S. Hantush, Tables of the Function $W(u, \beta)$, Professional Paper 104, Research and Development Division, New Mexico Institute of Mining and Technology, 1961.

u \ r/B	0	0.001	0.002	0.003	0.004	0.005	0.006	0.007	0.008	0.009	0.01
0	∞	14.0474	12.6611	11.8502	11.2748	10.8286	10.4640	10.1557	9.8887	9.6532	9.4425
.000001	13.2383	13.0031	12.4417	11.8153	11.2711	10.8283	10.4640	10.1557	9.8887		
.000002	12.5451	12.4240	12.1013	11.6716	11.2259	10.8174	10.4619	10.1554	9.8886	9.6532	
.000003	12.1397	12.0581	11.8322	11.5098	11.1462	10.7849	10.4509	10.1523	9.8879	9.6530	9.4425
.000004	11.8520	11.7905	11.6168	11.3597	11.0555	10.7374	10.4291	10.1436	9.8849	9.6521	9.4472
.000005	11.6289	11.5795	11.4384	11.2248	10.9642	10.6822	10.3993	10.1290	9.8786	9.6496	9.4413
.000006	11.4465	11.4053	11.2866	11.1040	10.8764	10.6240	10.3640	10.1094	9.8686	9.6450	9.4394
.000007	11.2924	11.2570	11.1545	10.9951	10.7933	10.5652	10.3255	10.0862	9.8555	9.6382	9.4361
.000008	11.1589	11.1279	11.0377	10.8962	10.7151	10.5072	10.2854	10.0602	9.8398	9.6292	9.4313
.000009	11.0411	11.0135	10.9330	10.8059	10.6416	10.4508	10.2446	10.0324	9.8219	9.6182	9.4251
.00001	10.9357	10.9109	10.8382	10.7228	10.5725	10.3963	10.2038	10.0034	9.8024	9.6059	9.4176
.00002	10.2426	10.2301	10.1932	10.1332	10.0522	9.9530	9.8386	9.7126	9.5781	9.4383	9.2961
.00003	9.8371	9.8288	9.8041	9.7635	9.7081	9.6392	9.5583	9.4671	9.3674	9.2611	9.1499
.00004	9.5495	9.5432	9.5246	9.4940	9.4520	9.3992	9.3366	9.2653	9.1863	9.1009	9.0102
.00005	9.3263	9.3213	9.3064	9.2818	9.2480	9.2052	9.1542	9.0957	9.0304	8.9591	8.8827
.00006	9.1440	9.1398	9.1274	9.1069	9.0785	9.0426	8.9996	8.9500	8.8943	8.8332	8.7673
.00007	8.9899	8.9863	8.9756	8.9580	8.9336	8.9027	8.8654	8.8224	8.7739	8.7204	8.6625
.00008	8.8563	8.8532	8.8439	8.8284	8.8070	8.7798	8.7470	8.7090	8.6661	8.6186	8.5669
.00009	8.7386	8.7358	8.7275	8.7138	8.6947	8.6703	8.6411	8.6071	8.5686	8.5258	8.4792
.0001	8.6332	8.6308	8.6233	8.6109	8.5937	8.5717	8.5453	8.5145	8.4796	8.4407	8.3983
.0002	7.9402	7.9390	7.9352	7.9290	7.9203	7.9092	7.8958	7.8800	7.8619	7.8416	7.8192
.0003	7.5348	7.5340	7.5315	7.5274	7.5216	7.5141	7.5051	7.4945	7.4823	7.4686	7.4534
.0004	7.2472	7.2466	7.2447	7.2416	7.2373	7.2317	7.2249	7.2169	7.2078	7.1974	7.1859
.0005	7.0242	7.0237	7.0222	7.0197	7.0163	7.0118	7.0063	6.9999	6.9926	6.9843	6.9750
.0006	6.8420	6.8416	6.8403	6.8383	6.8353	6.8316	6.8271	6.8218	6.8156	6.8086	6.8009
.0007	6.6879	6.6876	6.6865	6.6848	6.6823	6.6790	6.6752	6.6706	6.6653	6.6594	6.6527
.0008	6.5545	6.5542	6.5532	6.5517	6.5495	6.5467	6.5433	6.5393	6.5347	6.5295	6.5237
.0009	6.4368	6.4365	6.4357	6.4344	6.4324	6.4299	6.4269	6.4233	6.4192	6.4146	6.4094
.001	6.3315	6.3313	6.3305	6.3293	6.3276	6.3253	6.3226	6.3194	6.3157	6.3115	6.3069
.002	5.6394	5.6393	5.6389	5.6383	5.6374	5.6363	5.6350	5.6334	5.6315	5.6294	5.6271
.003	5.2349	5.2348	5.2346	5.2342	5.2336	5.2329	5.2320	5.2310	5.2297	5.2283	5.2267
.004	4.9482	4.9482	4.9480	4.9477	4.9472	4.9467	4.9460	4.9453	4.9443	4.9433	4.9421
.005	4.7261	4.7260	4.7259	4.7256	4.7253	4.7249	4.7244	4.7237	4.7230	4.7222	4.7212
.006	4.5448	4.5448	4.5447	4.5444	4.5441	4.5438	4.5433	4.5428	4.5422	4.5415	4.5407
.007	4.3916	4.3916	4.3915	4.3913	4.3910	4.3908	4.3904	4.3899	4.3894	4.3888	4.3882
.008	4.2591	4.2591	4.2590	4.2588	4.2586	4.2583	4.2580	4.2576	4.2572	4.2567	4.2561
.009	4.1423	4.1423	4.1422	4.1420	4.1418	4.1416	4.1413	4.1410	4.1406	4.1401	4.1396
.01	4.0379	4.0379	4.0378	4.0377	4.0375	4.0373	4.0371	4.0368	4.0364	4.0360	4.0356
.02	3.3547	3.3547	3.3547	3.3546	3.3545	3.3544	3.3543	3.3542	3.3540	3.3538	3.3536
.03	2.9591	2.9591	2.9591	2.9591	2.9590	2.9590	2.9589	2.9588	2.9587	2.9585	2.9584
.04	2.6813	2.6813	2.6812	2.6812	2.6811	2.6811	2.6810	2.6810	2.6809	2.6808	2.6807
.05	2.4679	2.4679	2.4679	2.4679	2.4678	2.4678	2.4678	2.4677	2.4676	2.4675	2.4675
.06	2.2953	2.2953	2.2953	2.2953	2.2952	2.2952	2.2952	2.2952	2.2951	2.2950	2.2950
.07	2.1508	2.1508	2.1508	2.1508	2.1508	2.1508	2.1507	2.1507	2.1507	2.1506	2.1506
.08	2.0269	2.0269	2.0269	2.0269	2.0269	2.0269	2.0269	2.0268	2.0268	2.0268	2.0267
.09	1.9187	1.9187	1.9187	1.9187	1.9187	1.9187	1.9187	1.9186	1.9186	1.9186	1.9185
.1	1.8229	1.8229	1.8229	1.8229	1.8229	1.8229	1.8229	1.8228	1.8228	1.8228	1.8227
.2	1.2227	1.2226	1.2226	1.2226	1.2226	1.2226	1.2226	1.2226	1.2226	1.2226	1.2226
.3	0.9057	0.9057	0.9057	0.9057	0.9057	0.9057	0.9057	0.9057	0.9056	0.9056	0.9056
.4	7024	7024	7024	7024	7024	7024	7024	7024	7024	7024	7024
.5	5598	5598	5598	5598	5598	5598	5598	5598	5598	5598	5598
.6	4544	4544	4544	4544	4544	4544	4544	4544	4544	4544	4544
.7	3738	3738	3738	3738	3738	3738	3738	3738	3738	3738	3738
.8	3106	3106	3106	3106	3106	3106	3106	3106	3106	3106	3106
.9	2602	2602	2602	2602	2602	2602	2602	2602	2602	2602	2602
1.0	0.2194	0.2194	0.2194	0.2194	0.2194	0.2194	0.2194	0.2194	0.2194	0.2194	0.2194
2.0	489	489	489	489	489	489	489	489	489	489	489
3.0	130	130	130	130	130	130	130	130	130	130	130
4.0	38	38	38	38	38	38	38	38	38	38	38
5.0	11	11	11	11	11	11	11	11	11	11	11
6.0	4	4	4	4	4	4	4	4	4	4	4
7.0	1	1	1	1	1	1	1	1	1	1	1
8.0	0	0	0	0	0	0	0	0	0	0	0

Table 7.2 *(continued)*

u \ r/B	0.01	0.015	0.02	0.025	0.03	0.035	0.04	0.045	0.05
0	9.4425	8.6319	8.0569	7.6111	7.2471	6.9394	6.6731	6.4383	6.2285
.000001									
.000002									
.000003	9.4425								
.000004	9.4422								
.000005	9.4413								
.000006	9.4394								
.000007	9.4361	8.6319							
.000008	9.4313	8.6318							
.000009	9.4251	8.6316							
.00001	9.4176	8.6313	8.0569						
.00002	9.2961	8.6152	8.0558	7.6111	7.2471				
.00003	9.1499	8.5737	8.0483	7.6101	7.2470				
.00004	9.0102	8.5168	8.0320	7.6069	7.2465	6.9394	6.6731		
.00005	8.8827	8.4533	8.0080	7.6000	7.2450	6.9391	6.6730		
.00006	8.7673	8.3880	7.9786	7.5894	7.2419	6.9384	6.6729	6.4383	
.00007	8.6625	8.3233	7.9456	7.5754	7.2371	6.9370	6.6726	6.4382	6.2285
.00008	8.5669	8.2603	7.9105	7.5589	7.2305	6.9347	6.6719	6.4381	6.2284
.00009	8.4792	8.1996	7.8743	7.5402	7.2222	6.9316	6.6709	6.4378	6.2283
.0001	8.3983	8.1414	7.8375	7.5199	7.2122	6.9273	6.6693	6.4372	6.2282
.0002	7.8192	7.6780	7.4972	7.2898	7.0685	6.8439	6.6242	6.4143	6.2173
.0003	7.4534	7.3562	7.2281	7.0759	6.9068	6.7276	6.5444	6.3623	6.1848
.0004	7.1859	7.1119	7.0128	6.8929	6.7567	6.6088	6.4538	6.2955	6.1373
.0005	6.9750	6.9152	6.8346	6.7357	6.6219	6.4964	6.3626	6.2236	6.0821
.0006	6.8009	6.7508	6.6828	6.5988	6.5011	6.3923	6.2748	6.1512	6.0239
.0007	6.6527	6.6096	6.5508	6.4777	6.3923	6.2962	6.1917	6.0807	5.9652
.0008	6.5237	6.4858	6.4340	6.3695	6.2935	6.2076	6.1136	6.0129	5.9073
.0009	6.4094	6.3757	6.3294	6.2716	6.2032	6.1256	6.0401	5.9481	5.8509
.001	6.3069	6.2765	6.2347	6.1823	6.1202	6.0494	5.9711	5.8864	5.7965
.002	5.6271	5.6118	5.5907	5.5638	5.5314	5.4939	5.4516	5.4047	5.3538
.003	5.2267	5.2166	5.2025	5.1845	5.1627	5.1373	5.1084	5.0762	5.0408
.004	4.9421	4.9345	4.9240	4.9105	4.8941	4.8749	4.8530	4.8286	4.8016
.005	4.7212	4.7152	4.7068	4.6960	4.6829	4.6675	4.6499	4.6302	4.6084
.006	4.5407	4.5357	4.5287	4.5197	4.5088	4.4960	4.4814	4.4649	4.4467
.007	4.3882	4.3839	4.3779	4.3702	4.3609	4.3500	4.3374	4.3233	4.3077
.008	4.2561	4.2524	4.2471	4.2404	4.2323	4.2228	4.2118	4.1994	4.1857
.009	4.1396	4.1363	4.1317	4.1258	4.1186	4.1101	4.1004	4.0894	4.0772
.01	4.0356	4.0326	4.0285	4.0231	4.0167	4.0091	4.0003	3.9905	3.9795
.02	3.3536	3.3521	3.3502	3.3476	3.3444	3.3408	3.3365	3.3317	3.3264
.03	2.9584	2.9575	2.9562	2.9545	2.9523	2.9501	2.9474	2.9444	2.9409
.04	2.6807	2.6800	2.6791	2.6779	2.6765	2.6747	2.6727	2.6705	2.6680
.05	2.4675	2.4670	2.4662	2.4653	2.4642	2.4628	2.4613	2.4595	2.4576
.06	2.2950	2.2945	2.2940	2.2932	2.2923	2.2912	2.2900	2.2885	2.2870
.07	2.1506	2.1502	2.1497	2.1491	2.1483	2.1474	2.1464	2.1452	2.1439
.08	2.0267	2.0264	2.0260	2.0255	2.0248	2.0240	2.0231	2.0221	2.0210
.09	1.9165	1.9183	1.9179	1.9174	1.9169	1.9162	1.9154	1.9146	1.9136
.1	1.8227	1.8225	1.8222	1.8218	1.8213	1.8207	1.8200	1.8193	1.8184
.2	1.2226	1.2225	1.2224	1.2222	1.2220	1.2218	1.2215	1.2212	1.2209
.3	0.9056	0.9056	0.9055	0.9054	0.9053	0.9052	0.9050	0.9049	0.9047
.4	7024	7023	7023	7022	7022	7021	7020	7019	7018
.5	5598	5597	5597	5597	5596	5596	5595	5594	5594
.6	4544	4544	4543	4543	4543	4542	4542	4542	4541
.7	3738	3738	3737	3737	3737	3737	3736	3736	3735
.8	3106	3106	3106	3106	3105	3105	3105	3105	3104
.9	2602	2602	2602	2602	2601	2601	2601	2601	2601
1.0	0.2194	0.2194	0.2194	0.2194	0.2193	0.2193	0.2193	0.2193	0.2193
2.0	489	489	489	489	489	489	489	489	489
3.0	130	130	130	130	130	130	130	130	130
4.0	38	38	38	38	38	38	38	38	38
5.0	11	11	11	11	11	11	11	11	11
6.0	4	4	4	4	4	4	4	4	4
7.0	1	1	1	1	1	1	1	1	1
8.0	0	0	0	0	0	0	0	0	0

0.055	0.06	0.065	0.07	0.075	0.08	0.085	0.09	0.095	0.10
6.0388	5.8658	5.7067	5.5596	5.4228	5.2950	5.1750.	5.0620	4.9553	4.8541
6.0388	5.8652	5.7067	5.5596	5.4228	5.2950				
6.0338	5.8637	5.7059	5.5593	5.4227	5.2949	5.1750	5.0620	4.9553	
6.0145	5.8527	5.6999	5.5562	5.4212	5.2942	5.1747	5.0619	4.9552	4.8541
5.9818	5.8309	5.6860	5.5476	5.4160	5.2912	5.1730	5.0610	4.9547	4.8539
5.9406	5.8011	5.6648	5.5330	5.4062	5.2848	5.1689	5.0585	4.9532	4.8530
5.8948	5.7658	5.6383	5.5134	5.3921	5.2749	5.1621	5.0539	4.9502	4.8510
5.8468	5.7274	5.6081	5.4902	5.3745	5.2618	5.1526	5.0471	4.9454	4.8478
5.7982	5.6873	5.5755	5.4642	5.3542	5.2461	5.1406	5.0381	4.9388	4.8430
5.7500	5.6465	5.5416	5.4364	5.3317	5.2282	5.1266	5.0272	4.9306	4.8368
5.7026	5.6058	5.5071	5.4075	5.3078	5.2087	5.1109	5.0133	4.9208	4.8292
5.2991	5.2411	5.1803	5.1170	5.0517	4.9848	4.9166	4.8475	4.7778	4.7079
5.0025	4.9615	4.9180	4.8722	4.8243	4.7746	4.7234	4.6707	4.6169	4.5622
4.7722	4.7406	4.7068	4.6710	4.6335	4.5942	4.5533	4.5111	4.4676	4.4230
4.5846	4.5590	4.5314	4.5022	4.4713	4.4389	4.4050	4.3699	4.3335	4.2960
4.4267	4.4051	4.3819	4.3573	4.3311	4.3036	4.2747	4.2446	4.2134	4.1812
4.2905	4.2719	4.2518	4.2305	4.2078	4.1839	4.1588	4.1326	4.1053	4.0771
4.1707	4.1544	4.1368	4.1180	4.0980	4.0769	4.0547	4.0315	4.0073	3.9822
4.0638	4.0493	4.0336	4.0169	3.9991	3.9802	3.9603	3.9395	3.9178	3.8952
3.9675	3.9544	3.9403	3.9252	3.9091	3.8920	3.8741	3.8552	3.8356	3.8150
3.3205	3.3141	3.3071	3.2997	3.2917	3.2832	3.2742	3.2647	3.2547	3.2442
2.9370	2.9329	2.9284	2.9235	2.9183	2.9127	2.9069	2.9007	2.8941	2.8873
2.6652	2.6622	2.6589	2.6553	2.6515	2.6475	2.6432	2.6386	2.6338	2.6288
2.4554	2.4531	2.4505	2.4478	2.4448	2.4416	2.4383	2.4347	2.4310	2.4271
2.2852	2.2833	2.2812	2.2790	2.2766	2.2740	2.2713	2.2684	2.2654	2.2622
2.1424	2.1408	2.1391	2.1372	2.1352	2.1331	2.1308	2.1284	2.1258	2.1232
2.0198	2.0184	2.0169	2.0153	2.0136	2.0118	2.0099	2.0078	2.0056	2.0034
1.9125	1.9114	1.9101	1.9087	1.9072	1.9056	1.9040	1.9022	1.9003	1.8983
1.8175	1.8164	1.8153	1.8141	1.8128	1.8114	1.8099	1.8084	1.8067	1.8050
1.2205	1.2201	1.2196	1.2192	1.2186	1.2181	1.2175	1.2168	1.2162	1.2155
0.9045	0.9043	0.9040	0.9038	0.9035	0.9032	0.9029	0.9025	0.9022	0.9018
7016	7015	7014	7012	7010	7008	7006	7004	7002	7000
5593	5592	559i	5590	5588	5587	5586	5584	5583	5581
4540	4540	4539	4538	4537	4536	4535	4534	4533	4532
3735	3734	3734	3733	3733	3732	3732	3731	3730	3729
3104	3104	3103	3103	3102	3102	3101	3101	3100	3100
2600	2600	2600	2599	2599	2599	2598	2598	2597	2597
0.2193	0.2192	0.2192	0.2192	0.2191	0.2191	0.2191	0.2191	0.2190	0.2190
489	489	489	489	489	489	489	489	488	488
130	130	130	130	130	130	130	130	130	130
38	39	38	38	38	38	38	38	38	38
11	11	11	11	11	11	11	11	11	11
4	4	4	4	4	4	4	4	4	4
1	1	1	1	1	1	1	1	1	1
0	0	0	0	0	0	0	0	0	0

Table 7.2—(continued)

Columns are values of r/B; rows are values of u. (Blank cells indicate no value printed.)

u	0.1	0.15	0.2	0.25	0.3	0.35	0.4	0.45	0.5	0.55	0.6	0.65	0.7	0.75	0.8	0.85	0.9	0.95	1.0
0	4.8541	4.0601	3.5054	3.0830	2.7449	2.4654	2.2291	2.0258	1.8488	1.6981	1.5550	1.4317	1.3210	1.2212	1.1307	1.0485	0.9735	0.9049	0.8420
.0001																			
.0002																			
.0003	4.8541																		
.0004	4.8539																		
.0005	4.8530																		
.0006	4.8510	4.0601																	
.0007	4.8478	4.0600																	
.0008	4.8430	4.0599																	
.0009	4.8368	4.0598																	
.001	4.8292	4.0595	3.5054	3.0830															
.002	4.7079	4.0435	3.5043	3.0821				2.0258											
.003	4.5622	4.0092	3.4806	3.0788	2.7449			2.0257											
.004	4.4230	3.9551	3.4567	3.0719	2.7444	2.4654	2.2291	2.0256	1.8488										
.005	4.2960	3.8821			2.7428	2.4651	2.2290	2.0253	1.8487										
.006	4.1812	3.8384	3.4274	3.0614	2.7398	2.4644	2.2289												
.007	4.0771	3.7529	3.3947	3.0476	2.7350	2.4630	2.2286												
.008	3.9822	3.6903	3.3598	3.0311	2.7284	2.4608	2.2279												
.009	3.8952	3.6302	3.3329	3.0126	2.7202	2.4576	2.2269												
.01	3.8150	3.5725	3.2875	2.9925	2.7104	2.4534	2.2253	2.0248	1.8486	1.6931	1.5550	1.4317	1.3210	1.2212	1.1307	1.0485	0.9735	0.9049	0.8420
.02	3.2442	3.1158	2.9521	2.7658	2.5688	2.3713	2.1809	2.0023	1.8379	1.6883	1.5530	1.4309	1.3207	1.2210	1.1306	1.0484	0.9733	0.9048	0.8418
.03	2.8873	2.8017	2.6896	2.5758	2.4110	2.2578	2.1031	1.9515	1.8062	1.6695	1.5423	1.4251	1.3177	1.2195	1.1299	1.0481	0.9724	0.9044	0.8409
.04	2.6328	2.5655	2.4816	2.3802	2.2661	2.1431	2.0155	1.8869	1.7603	1.6379	1.5213	1.4117	1.3094	1.2146	1.1270	1.0465	0.9700	0.9029	0.8391
.05	2.4271	2.3776	2.3110	2.2299	2.1371	2.0356	1.9283	1.8181	1.7075	1.5985	1.4927	1.3914	1.2955	1.2052	1.1210	1.0426	0.9657	0.9001	0.8360
.06	2.2622	2.2218	2.1673	2.1002	2.0227	1.9369	1.8452	1.7497	1.6524	1.5551	1.4593	1.3663	1.2770	1.1919	1.1116	1.0362	0.9593	0.8956	0.8316
.07	2.1232	2.0894	2.0435	1.9867	1.9206	1.8469	1.7673	1.6835	1.5973	1.5101	1.4232	1.3380	1.2551	1.1754	1.0993	1.0272	0.9510	0.8895	0.8259
.08	2.0034	1.9745	1.9351	1.8861	1.8290	1.7646	1.6947	1.6206	1.5436	1.4650	1.3860	1.3078	1.2310	1.1564	1.0847	1.0161	0.9411	0.8819	
.09	1.8983	1.8732	1.8389	1.7961	1.7460	1.6892	1.6272	1.5609	1.4918	1.4206	1.3486	1.2766	1.2054	1.1358	1.0682	1.0032			
.1	1.8050	1.7829	1.7527	1.7149	1.6704	1.6198	1.5644	1.5048	1.4222	1.3774	1.3115	1.2451	1.1791	1.1140	1.0505	0.9890	0.9297	0.8730	0.8190
.2	1.2155	1.2066	1.1944	1.1789	1.1602	1.1387	1.1145	1.0879	1.0592	1.0286	0.9964	0.9629	0.9284	0.8932	0.8575	0.8216	0.7857	0.7501	0.7148
.3	0.9018	0.8969	0.8902	0.8817	0.8713	0.8593	0.8457	0.8306	0.8142	0.7964	0.7775	0.7577	0.7369	0.7154	0.6932	0.6706	0.6476	0.6244	0.6010
.4	0.7000	0.6969	0.6927	0.6874	0.6809	0.6733	0.6647	0.6551	0.6446	0.6332	0.6209	0.6080	0.5943	0.5801	0.5653	0.5501	0.5345	0.5186	0.5024
.5	0.5581	0.5561	0.5532	0.5496	0.5453	0.5402	0.5344	0.5278	0.5206	0.5128	0.5044	0.4955	0.4860	0.4761	0.4658	0.4550	0.4440	0.4326	0.4210
.6	0.4532	0.4518	0.4498	0.4472	0.4441	0.4405	0.4364	0.4317	0.4266	0.4210	0.4150	0.4086	0.4018	0.3946	0.3871	0.3793	0.3712	0.3629	0.3543
.7	0.3729	0.3719	0.3704	0.3585	0.3663	0.3636	0.3606	0.3572	0.3534	0.3493	0.3449	0.3401	0.3351	0.3297	0.3242	0.3183	0.3123	0.3060	0.2996
.8	0.3100	0.3092	0.3081	0.3067	0.3052	0.3030	0.3008	0.2982	0.2953	0.2922	0.2889	0.2853	0.2815	0.2774	0.2732	0.2687	0.2641	0.2592	0.2543
.9	0.2597	0.2591	0.2583	0.2572	0.2559	0.2544	0.2527	0.2507	0.2485	0.2461	0.2436	0.2408	0.2378	0.2347	0.2314	0.2280	0.2244	0.2207	0.2168
1.0	0.2190	0.2186	0.2179	0.2171	0.2161	0.2149	0.2135	0.2120	0.2103	0.2085	0.2065	0.2043	0.2020	0.1995	0.1970	0.1943	0.1914	0.1885	0.1855
2.0	0.0488	0.0488	0.0487	0.0486	0.0485	0.0484	0.0482	0.0480	0.0477	0.0475	0.0473	0.0470	0.0467	0.0463	0.0460	0.0456	0.0452	0.0448	0.0444
3.0	0.0130	0.0130	0.0130	0.0130	0.0130	0.0130	0.0129	0.0129	0.0128	0.0128	0.0127	0.0127	0.0126	0.0125	0.0125	0.0124	0.0123	0.0123	0.0122
4.0	0.0038	0.0038	0.0038	0.0038	0.0038	0.0038	0.0038	0.0037	0.0037	0.0037	0.0037	0.0037	0.0037	0.0037	0.0037	0.0036	0.0036	0.0036	0.0036
5.0	0.0011	0.0011	0.0011	0.0011	0.0011	0.0011	0.0011	0.0011	0.0011	0.0011	0.0011	0.0011	0.0011	0.0011	0.0011	0.0011	0.0011	0.0011	0.0011
6.0	0.0004	0.0004	0.0004	0.0004	0.0004	0.0004	0.0004	0.0004	0.0004	0.0004	0.0004	0.0004	0.0004	0.0004	0.0004	0.0004	0.0004	0.0004	0.0004
7.0	0.0001	0.0001	0.0001	0.0001	0.0001	0.0001	0.0001	0.0001	0.0001	0.0001	0.0001	0.0001	0.0001	0.0001	0.0001	0.0001	0.0001	0.0001	0.0001
8.0	0.0000	0.0000	0.0000	0.0000	0.0000	0.0000	0.0000	0.0000	0.0000	0.0000	0.0000	0.0000	0.0000	0.0000	0.0000	0.0000	0.0000	0.0000	0.0000

u \ r/B	1.0	1.5	2.0	2.5	3.0	3.5	4.0	4.5	5.0	6.0	7.0	8.0	9.0
0	0.8420	0.4276	0.2278	0.1247	0.0695	0.0392	0.0223	0.0128	0.0074	0.0025	0.0008	0.0003	0.0001
.01	0.8420												
.02	8418												
.03	8409												
.04													
.05													
.06	8391												
.07	8360	0.4276											
.08	8316	4275											
.09	8259	4274											
.1	0.8190	0.4271	0.2278										
.2	7148	4135	2268	0.1247	0.0695								
.3	6010	3812	2211	1240	694								
.4	5024	3411	2096	1217	691	0.0392							
.5	4210	3007	1944	1174	681	390	0.0223						
.6	3543	2630	1774	1112	664	386	222	0.0128					
.7	2996	2292	1602	1040	639	379	221	127					
.8	2543	1994	1436	961	607	368	218	127	0.0074				
.9	2168	1734	1281	881	572	354	213	125	73				
1.0	0.1855	0.1509	0.1139	0.0803	0.0534	0.0338	0.0207	0.0123	0.0073	0.0025	0.0008	0.0003	0.0001
2.0	444	394	335	271	210	156	112	77	51	21	6	2	0
3.0	122	112	100	86	71	57	45	34	25	12	3	1	
4.0	36	34	31	27	24	20	16	13	10	6	1	0	
5.0	11	10	10	9	8	7	6	5	4	2			
6.0	4	3	3	3	3	2	2	2	2	1	1		
7.0	1	1	1	1	1	1	1	1	1	0	0		
8.0	0	0	0	0	0	0	0	0	0				

those in Figure 7.7 and as done previously in Section 2.7, the contact point of $W(u, \beta)$, $1/u$, s, and r^2/t is obtained. Then, T and S can be calculated using the formulas:

$$T = \frac{QW(u, \beta)}{4\pi s} \tag{7.5-2}$$

and

$$S = 4\pi u(t/r^2) \tag{7.5-3}$$

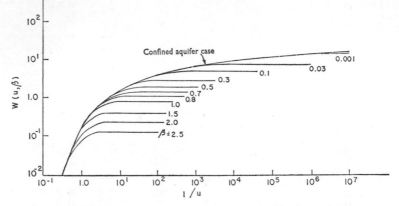

Figure 7.7 Leaky aquifer type curves (after Walton, 1960).

Formula (7.5-1) and the method of calculating S and T outlined here are subject to some restrictions since we have assumed that:

(1) the aquifer is infinite in areal extent, uniformly thick, homogeneous, and isotropic,
(2) the aquifer is located over an impermeable bed and below a leaky formation,
(3) the storage coefficient is constant,
(4) the well diameter is small compared to the dimensions of the aquifer,
(5) the well fully penetrates the aquifer, and
(6) the flow is horizontal in the aquifer and vertical in the leaky formation.

Even with all these restrictions formula (7.5-1) is extremely useful in calculating the formation constants of a leaky aquifer.

7.6 Application of Theis' Theory to Unconfined Aquifers

The theories discussed in the preceding sections of this chapter are applicable to fully confined or leaky aquifers. Formula (7.1-7) for confined aquifers is, according to Boulton,[5] a good approximation to obtain the drawdown in an unconfined aquifer, if certain conditions are satisfied. Using Boulton's theory, Walton[6] has indicated that formula (7.1-7) can be applied to unconfined aquifers for times larger than

$$t_{nc} = 37.4Rb/K \qquad 0.2b < r < 6b \qquad (7.6\text{-}1)$$

where t_{nc} = time in days after which formula (7.1-7) can be applied, R = specific yield, b = saturated thickness in feet, and K = hydraulic conductivity in gallons per day per ft².

Jacob[7] also carried studies for the purpose of adapting equation (7.1-7) to the case of unconfined aquifers. He found a correspondence that permits one to apply formula (7.1-7) to unconfined aquifers and knowing the drawdown s' given by the formula, the real drawdown s can be obtained and viceversa. The Jacob correspondence principle states that

$$s' = s - s^2/2b \qquad (7.6\text{-}2)$$

where s' = drawdown in equivalent confined aquifer, s = drawdown observed in the unconfined aquifer, and b = saturated thickness.

Using Jacob's Principle and equation (7.1-7) the formation constants for an unconfined aquifer can be obtained using the method of Section 7.2.

7.7 Time Needed for Stabilization of the Drawdown Cone

A considerable amount of time may be required in some instances in order for an equilibrium state to be reached. The time needed to approximately reach an equilibrium state may be obtained by means of the Foley-Walton-Drescher equation[8] which states that

$$t_{eq} = a^2S/[112T\epsilon \log (2a/r)^2] \qquad (7.7\text{-}1)$$

---

[5] Boulton, N. S. (1954) *The drawdown of the water table under nonsteady conditions near a pumped well in an unconfined formation*, Proc. British Institute of Civil Engineers, vol. 3, pt. 3.

[6] Walton, W. C. (1962) *Selected analytical methods for well and aquifer evaluation*, Illinois State Water Survey Bulletin #49.

[7] Jacob, C. E. (1944) *Notes on determining permeability by pumping tests under water-table conditions*, U.S.G.S. mimeographed report.

[8] Foley, F. C., Walton, W. C., and Drescher, W. J. (1953) *Groundwater Conditions in the Milwaukee Waukesha area, Wisconsin*, U.S.G.S. Water Supply Paper 1229.

where t_{eq} = time needed to reach an approximate equilibrium state (in years), a = distance from the pumping well to the zone of recharge (in feet), r = distance from the pumping well to the observation point (in feet), S = storage coefficient (nil dimensional), T = transmissivity (in gallons per day/ft²), and ϵ = deviation from absolute equilibrium.

7.8   Step-drawdown Tests: Practical Cases

In order to quantify the hydrodynamic properties of an aquifer certain tests are needed. We have already discussed pumping tests and have been able to determine the amount of water in storage and the aquifer's transmissivity. Another type of test which is very useful is the so called step-drawdown test. This test permits one to evaluate the conditions of flow into a well, i.e., whether the well is rapidly deteriorating or not. They also allow a calculation of well losses.

Step-drawdown tests are known in the petroleum industry as "multiple rate flow tests." Such tests consist in pumping an amount Q_1 of water or oil as the case may be and changing this amount to Q_2 after a time Δt has elapsed. The steps can be continued i times, each change representing one

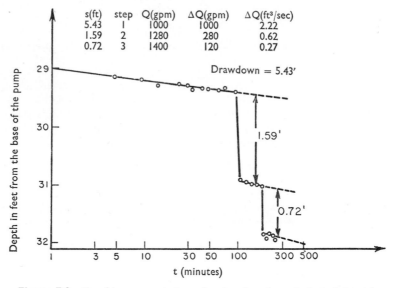

Figure 7.8   Graphic representation of a step-drawdown test carried out by the Illinois State Water Survey near Granite City, Illinois (taken from Bruin and Hudson, 1955).

step. Figure 7.8 taken from Bruin and Hudson[9] shows a typical step-drawdown test. The information supplied by these tests together with stratigraphic and geophysical data allows a detailed evaluation of conditions at the well. Also it furnishes a picture of aquifer behavior under pumpage.

The theoretical basis for the step-drawdown test is rather simple. One can individually consider each interval of time where an amount Q_i is pumped and then apply the continuous form of the superposition principle (see Carslaw and Jaeger[10]).

In Section 6.11 the effect that a well has on the drawdown near it was discussed. At that time, it was shown how Jacob and Rorabaugh developed theories to account for this effect. There, we also indicated the existence of a quantity C that Walton has used successfully as an indicator of well deterioration. Step-drawdown tests allow a determination of this quantity. To illustrate how this is done a step-drawdown test carried in a well near Granite City, Illinois by the Illinois State Water Survey[9] will be used. Results of this test are presented in tabular form on Table 7.3 and in graphical form in Figure 7.8 Using this data C can be calculated using Jacob's equation[11]

$$C = \frac{(\Delta s_i/\Delta Q_i) - (\Delta s_{i-1}/\Delta Q_{i-1})}{\Delta Q_{i-1} + \Delta Q_i} \qquad (7.8\text{-}1)$$

where the Δs represent increments in the drawdown caused by increments ΔQ in the discharge. Consequently, for steps 1 and 2

$$C = \frac{(\Delta s_2/\Delta Q_2) - (\Delta s_1/\Delta Q_1)}{\Delta Q_1 + \Delta Q_2} \qquad (7.8\text{-}2)$$

and for steps 2 and 3

$$C = \frac{(\Delta s_3/\Delta Q_3) - (\Delta s_2/\Delta Q_2)}{\Delta Q_2 + \Delta Q_3} \qquad (7.8\text{-}3)$$

It must be noted that all the equations above assume that C is a constant throughout the test. This is not the case in many instances.

To estimate well losses from the above data (see Section 6.11) one first calculates the average value of C, in this case $C = 0.16$ sec^2/ft^5. Then, using

[9] Bruin, J. and Hudson Jr., J. E. (1955) *Selected Methods for Pumping Test Analysis*, Illinois State Water Survey, Report of Investigation 25.

[10] Carslaw, H. S. and Jaeger, J. C. (1959) *Conduction of Heat in Solids*, Oxford at the Clarendon Press.

[11] Jacob, C. E. (1947) *Drawdown test to determine effective radius of artesian well*, Trans. Am. Soc. Civil Engrs., vol 112, p. 1049.

Table 7.3 Data from a step-drawdown
test near Granite City, Illinois (from Bruin
and Hudson, 1955).

Time	Feet to the water	$\dfrac{Q}{(\text{gpm})}$
9:13 a.m.	23.95	0
9:23	23.95	0
9:45	pumping started	
9:50	29.13	1000
9:55	29.23	1000
10:00	29.30	1000
10:10	29.30	1000
10:15	29.32	1000
10:20	29.37	1000
10:30	29.35	1000
10:35	29.36	1000
10:52	29.40	1000
11:00	29.39	1000
11:25	29.43	1000
11:30	30.97	1280
11:50	31.01	1280
12:00	31.02	1280
12:15 p.m.	31.04	1280
12:30	31.08	1280
12:45	31.08	1280
12:50	31.90	1400
1:00	31.95	1400
1:15	31.90	1400
1:30	31.90	1400
1:40	31.93	1400

equation (6.11-2) one can calculate the losses. For example, when $Q = 1400$ gpm the well losses will be

$$CQ^2 = 7.74 \text{ ft}$$

and this is a considerable fraction of the total drawdown (over 10% in the case considered above).

Exercises

1. Examine the convergence of the series expansion of the Well Function $W(u)$ when (a) $u > 0.02$; (b) $0.01 \le u \le 0.02$; (c) $u < 0.01$.

2. Starting from formula (7.1-11) for confined aquifers prove that for an unconfined aquifer

$$h_1{}^2 - h_2{}^2 = \frac{2.3Q}{2\pi K} \log \frac{t_2}{t_1}$$

when $r_1 = r_2$.

3. In a confined aquifer the drawdown caused by pumping a well is 0.35 meters at a distance of 30 meters from the well after two hours from the beginning of pumping. How long a time will elapse before drawdowns of 0.35 meters are observed at 100 meters from the well?

4. A well in a confined aquifer 80 meters thick produces 1.8×10^6 liters per day of water. At a distance of 15 meters, drawdowns of 1.53, 2.87, 4.40, and 5.50 meters are observed after 1, 3, 8, and 16 hours have elapsed from the beginning of pumping. At a distance of 30 meters the corresponding drawdowns are 0.33, 1.21, 2.36, and 3.30. Calculate S and T using Theis' method.

5. Repeat Problem 4 using Jacob's method.

6. Discuss the physical meaning of conditions (7.4-2), (7.4-3), and (7.4-4).

7. Under what conditions (physical and geologic) is formula (7.5-1) valid?

8. Prove that the average value of C is 0.16 sec²/ft⁵ for the step-drawdown test discussed in Section 7.8.

9. Using the results of the previous problem calculate what percent of the total drawdown is made up by well losses if Q is 1400 gallons per minute using (a) Jacob's formula; and (b) Rorabaugh's formula assuming that n is 1.5, 1.7, 1.9, 2.1, and 2.3.

PART 4

REGIONAL HYDROLOGY

CHAPTER 8

Regional Subsurface Hydrology

The utilization of water in a large scale in a given basin requires the exploration of all the water resources in the area and the evaluation of the different sources of water supply. These integral basin-wide studies are the concern of *Regional Hydrology*. In this chapter the aspects of Regional Hydrology dealing with subsurface resources will be discussed.

One can consider *Subsurface Regional Hydrology* as encompassing three parts. These are: the exploration phase, the drilling operation, and the evaluation of the subsurface aquifers as sources of water supply. Once the sources of water have been established, the practical aspects of water supply are the concern of a water-supply engineer and are not considered a part of Subsurface Regional Hydrology.

Before exploiting the groundwater resources of a basin, it is absolutely necessary to perform a geologic and hydrologic reconnaissance that would permit us to pinpoint with sufficient accuracy the position of the aquifers in the zone and allow a determination of the amount of water stored in these aquifers. These goals require the following studies:

1. A geologic study and mapping of the zone including the location of faults and fractures and the estimation of the porosity and permeability of the various formations
2. Location of existing water wells in the zone (producing or not) and construction of piezometric head contours.
3. Determination of the quality of water in the wells already drilled in the basin. Construction of quality contours.
4. Evaluation of the socioeconomic and agricultural conditions prevailing in the basin. This can be done either by dynamic programming or by classical techniques.

These studies can, and in many cases need be, complemented by geophysical and geochemical surveys which will allow us to better pinpoint those aquifers with sufficient water of sufficiently good quality for the use desired. The geophysical techniques will now be discussed. Geochemical surveys are the subject of Chapter 10.

The geophysical exploration techniques used to locate ground-water bodies are of four basic types:

1. Gravimetric Surveys
2. Electric Surveys
   (i) resistivity
   (ii) spontaneous potential
3. Radioactive Techniques (Gamma Rays)
4. Seismic Surveys
   (i) reflection
   (ii) refraction

Having carried out a complete geological-geophysical-geochemical survey of the basin (as needed), a hydrologist can evaluate the potential of using subsurface waters as a source for water supply. This is the most difficult and crucial point of the entire study.

Once the exploratory phase has been completed and a preliminary evaluation of the overall subsurface water resources has been carried, the hydrologist must proceed to determine where and how deep to drill the wells to be used for water supply. The techniques to be used in drilling will be determined by a knowledge of stratigraphic conditions at the point to be drilled.

Aside from the exploration and exploitation of the subsurface water resources, it is the responsibility of the regional hydrologist to be able to evaluate the potential of these resources not only looking at the present but trying to project to the required socioeconomic demands of the future.

8.1 Geophysical Exploration Techniques

One of the geophysical methods used in groundwater exploration is the gravimetric survey. The gravimeters used in this type of work are portable instruments that record the variations in the value of the vertical component of gravity between various points in an interlocking network.

Gravimetric surveys have been used (a) to determine anomalies caused by distorted formations, (b) to locate salt domes, and (c) to locate aquifers. The technique simply consists in measuring with the gravimeter the value of the vertical component of gravity in numerous points in the basin, correcting these values to standard conditions, and constructing a gravimetric map of the basin. This map is the tool the hydrologist uses in interpreting the gravimetric survey. A good survey can examine gravity differences as small as 10^{-4} cm/sec^2.

When gravity values are measured at points in a basin these values must be corrected to account for the presence of adjacent valleys and mountains, changes in elevation, horizontal changes in mass, changes in topographical configuration of the buried rock surface, horizontal mass discontinuities, and earth rotation. Once all these corrections are applied, values can be compared for all points in the interlocking-basin network. A more detailed discussion of the gravimetric survey can be found in most books on geophysical exploration.[1]

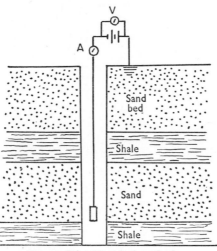

Figure 8.1   Electrical method.

A second geophysical technique of use in groundwater exploration is the electric survey. This method consists in determining the resistance that the various formations penetrated by a well give to the passage of an electric current. This resistance can be measured by slowly lowering one or more electrodes into the well and recording the resulting resistance-depth curve. Figure 8.1 illustrates a single electrode electric survey. Multiple electrodes are also used and preferred for petroleum exploration. However, the single-electrode technique is the most widely used by water contractors.

When using the electric method one expresses the resistance in terms of "resistivity" which is the reciprocal of electrical conductivity. The units of resistivity are ohms·m²/m.

An electric survey is composed of several independent measurements (see Figure 8.2) which are usually carried out simultaneously. As shown on Figure

---

[1] Jacobs, J. A., Russell, R. D., and Wilson, J. Tuzo (1959) *Physics and Geology*, McGraw-Hill Co., N.Y. New York.

8.2 some of these are spontaneous potential and apparent resistivity. It is obvious from this figure that the spontaneous potential of shale is more positive than that of a fresh-water sand, and the spontaneous potential of this last one is more positive than that of a brackish-water sand. The apparent resistivity also produces characteristic values for each of the penetrated formations.

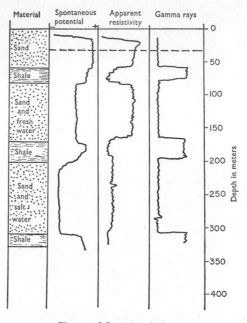

Figure 8.2   Electric log.

An electric log can be used for various purposes:

1. to determine the physical properties of the aquifers penetrated,
2. to relate formations penetrated by different wells,
3. to serve as an indicator of groundwater quality, and
4. to locate salt or connate water bodies.

Resistivity differences between different formations are reasonably large and in most cases not hard to distinguish. For example:

Formation	Resistivity
	ohms·m²/m
Shale	0–100
Shale and Saturated Sands	100–500
Dry Sands	500–1500
Gravel	1500–4000

An important objective of some resistivity surveys is the location of salt water. To understand how this is done consider now Archie's equation

$$F = \frac{R_0}{R_w}$$

where F is the formation resistivity factor, R_0 is the resistivity of a saturated sand, and R_w is the resistivity of the formation water.

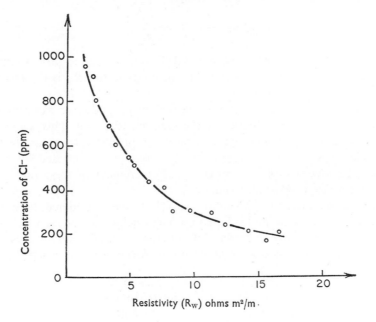

Figure 8.3 Concentration of Cl⁻ versus R_w for several samples taken from wells drilled in the same aquifer.

The factor F can be estimated for a given formation. The resistivity R_w can be measured by sampling waters from different wells penetrating the formation. If the Cl⁻ concentration in these samples is also measured we obtain a graph (see Figure 8.3) of Cl⁻ concentration versus R_w. Using Archie's equation this graph can be converted to a plot of Cl⁻ versus R_0. Once this standard plot is available Cl⁻ concentration can be calculated in the field by measuring R_0. This same type of analysis can be used to determine total dissolved solids instead of salinity (Cl⁻ concentration).

Some of the factors influencing these indirect water quality measurements are:

(a) Temperature variations at different points in the same aquifer affect resistivity values.

(b) At the boundary between two aquifers the Cl⁻ concentration and the amount of total dissolved solids will be a mixture of the respective concentrations in both aquifers.

(c) If the aquifer's sand contains a large amount of shale, the method can not be applied.

The measurement of spontaneous potential is related to resistivity by Patten's equation[2]

$$\Delta SP = -K \log \frac{R_m}{R_w}$$

where ΔSP is the spontaneous potential deflection, K is an empirical constant, R_w is the resistivity of the water in the formation, and R_m is the resistivity of the fluid in the borehole.

Another survey which is usually carried in groundwater exploration to obtain stratigraphic information is the Gamma Ray Log. This log consists in measuring the emission of gamma radiation produced by naturally radioactive elements present in the different formations penetrated by a well. Figure 8.2 shows a typical gamma log. Measurement of these rays can be done with a Geiger-Mueller Counter.

The last geophysical exploration survey to be considered here is the seismic survey. This method is based on the principle that shock waves travel through different earth materials at different speeds. These waves travel faster the denser the material making up the formation. For example:

Material	Speed of shock waves meters/second
Alluvium	300–600
Shale	2000–3000
Lutite	2000–4000
Salt domes	4000–5000
Granite	5000–6000

The shock waves are started by setting off an explosive charge in a small hole near the ground. The waves' arrival times are measured using a seismograph and the seismic velocities are computed using standard techniques as used in seismology. A more detailed discussion can be found in an article by Linehan and Keith.[3]

---

[2] Patten, E. P. and Bennett, Gordon (1963) *Application of Electrical and Radioactive Well Logging to Ground-water Hydrology*, USGS, WSP 1544-D.

[3] Linehan, D. and Keith, Scott (1948) *Seismic Reconnaissance for Ground Water Development*, Journal New England Water Works Association, vol. 63, n. 1.

8.2 Well Drilling Techniques

Once the exploratory phase has been completed and the well sites have been determined one can begin to drill. There are two basic drilling techniques which are used in the water industry. These are: rotary drilling and percussion. Also, wells are sometimes not drilled but dug, bored, or augered,

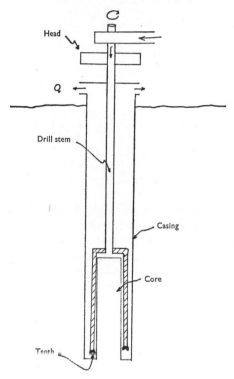

Figure 8.4 Rotary core drill.

however, these methods can be used only for relatively shallow wells (less than 30 meters deep).

The rotary method consists in abrading the rock by the rotation of a toothed or wheeled bit. An illustration of a rotary drill is the core drill (see Figure 8.4) which is simply a hollow bit possessing diamond abrasive teeth. The bit rotates and water is forced through the drill stem downward to remove cuttings. This technique is rather expensive. Figure 8.5b shows a typical rotary (rocker rock) bit.

The percussion technique involves letting a bit drop regularly against the formation to be drilled. The bit's impact breaks the formation slowly into small fragments. The speed of advance of a percussion drill is about four times less than that of a rotary tool. Also the maximum depth to which one

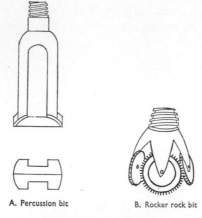

A. Percussion bit   B. Rocker rock bit

Figure 8.5   Drilling bits.

can drill by the percussion method is much less. However, the high cost of rotary drilling makes the percussion method preferential in some instances. Table 8.1 illustrates the advantages and disadvantages of both methods.

Once the well has been drilled it must be cased and completed. The portion of the well penetrating the aquifer to be exploited must possess screens to permit easy access of water and to prevent drastic pressure changes at the point of entry.

There are numerous types of water pumps, each useful under certain circumstances. There are revolving vertical shaft, jet, airlift, rotary positive displacement, and deep well centrifugal pumps (see Figure 8.6). Many factors must be considered before investing in a pump. Most important, the pump must be able to efficiently deliver the desired amount of water running under the specified conditions.

The deep well centrifugal pump is one of the most popular ones. It is a turbine type vertical centrifugal pump and consists of a discharge head, a column pipe and a pump assembly. The diameter of this pump is about 2 cm less than the well diameter and the pump is located about 5 m below the water level in the well. The capacity of such pumps is very high. For example, for a 25 cm diameter well the pump's capacity is about 200 cubic meters per hour and for larger diameters capacities in excess of 500–600 cubic meters per hour can be obtained. Also the deep well centrifugal pump is very efficient.

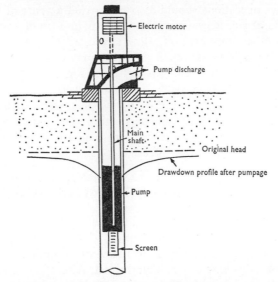

Figure 8.6 Deep well centrifugal pump.

Once the well is completed and all pumping and other tests have been run (see Part 3. Well Hydraulics) the well must be cleaned and disinfected thoroughly with a chlorine solution. Also all oil and grease must be removed from the casing pipe.

Table 8.1 Comparison of drilling techniques

	Rotary	Percussion
Drilling speed in sand and unconsolidated gravel formations	100–200 meters/day	20–30 meters/day
Drilling speed in consolidated and igneous formations	4–6 meters/day	1–2 meters/day
Maximum drilling depth	4000–5000 m	500 meters
Cost/meter	very high	low
Amount of water used in the drilling operation	large	small
Muds are used	yes	no
Drilling equipment	complex and costly	simple and cheap
Bits	Figure 8.5b	Figure 8.5a

Disinfection operations should also be carried as part of the regular job of well maintenance. Periodic maintenance is important and must be done to prevent well screen corrosion, eliminate incrustations, repair the pump, and perform any additional repair jobs that may be needed.

At times it is found that a well is not able to produce the desired yield of water. In many cases better yields can be achieved by acid treatment of the zone immediately adjacent to the well. The purpose of such treatment is to create fractures and increase the permeability of the region immediately adjacent to the well. The reader is referred to an article by Rowan where acid treatment is discussed in further detail.[4]

8.3   Regional Evaluation of Subsurface Water Resources

The groundwater resources of the aquifers in a given basin can be evaluated by various means. Using the principles of well hydraulics discussed in Chapters 6 and 7, the flow regime in the various aquifers can be evaluated and formation constants can be determined. Also the overall extraction of water from all aquifers in the basin can be optimized using techniques of well hydraulics.

It is also important to carry basin-wide evaluations of the water resources of an area. The simplest technique to accomplish this is the hydrologic inventory (see Chapter 2). To evaluate the groundwater reserves of a basin by this method one must begin by clearly and definitely limiting the basin. Once this is done one can calculate on the basis of field data from stations in the basin, the average precipitation, evaporation, and runoff. Also, the total pumpage from the basin must be computed from existing records and the interrelation between the basin and its neighbors must also be defined.

On the basis of this information the annual groundwater reserves of the basin can be computed using the formula

$$\bar{R} = \bar{P} - \overline{Ev} - \overline{Ru} - \overline{Pu} + \bar{I}$$

where

$\bar{R}$ = annual reserves,

$\bar{P}$ = average annual precipitation,

$\overline{Ev}$ = average annual evaporation,

$\overline{Ru}$ = average annual runoff,

$\overline{Pu}$ = annual pumpage, and

$\bar{I}$ = contributions from neighboring areas.

---

[4] Rowan, G. (1959) *Theory of Acid Treatments of Limestone Formations*, Journal Inst. Petr., vol. 45, n. 431.

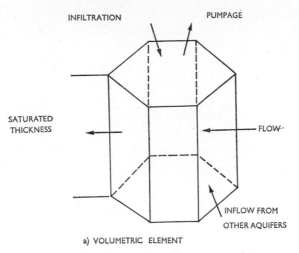

a) VOLUMETRIC ELEMENT

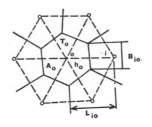

b) PLANAR ELEMENT

Figure 8.7 Finite elements.

If the quantity $\bar{R}$ is positive the water level of the aquifers or the piezo-metric surface must be rising. On the contrary, if for a given year $\bar{R}$ is negative this indicates an overpumpage of the aquifers in the zone and a lowering of the piezometric surface.

Hydrologic inventories are very useful especially for preliminary regional evaluations. Nonetheless, caution should be used in screening the data used in the inventory.

A more sophisticated analysis is the Cruickshank-Chávez Model[5] which is based on Darcy's law and the Principle of Conservation of Mass. This model is used to project piezometric levels to future situations. In this model the aquifer is considered to be divided into volumetric elements polygonal in shape. In each element (see Figure 8.7A) the occurrence of horizontal flow,

[5] Cruickshank, C., and Chávez-Guillén, R. (1969) *Modelo Matemático DAS para el Estudio del Comportamiento de Acuíferos*, Ingeniería Hidraúlica en México, vol. XXIII.

infiltration, pumpage, and contributions from neighboring elements are taken into account. The object of the model is simply to produce a balance between the water entering each volumetric element and the water leaving it.

Instead of using volumetric elements it is easier to use planar polygons as in Figure 8.7B, and define the following parameters:

T_i = transmissivity of element i,

T_o = transmissivity of element o,

$h_{i,k+1}$ = piezometric head of node i at the instant $k + 1$,

$h_{i,k}$ = piezometric head of node i at the instant k,

$h_{o,k+1}$ = piezometric head of node o at the instant $k + 1$,

$h_{o,k}$ = piezometric head of node o at the instant k,

L_{io} = distance between the centers of polygons i and o,

B_{io} = side length of the polygon corresponding to the side perpendicular to io,

n = number of polygons adjacent to o,

$Q_{o,k}$ = pumpage of polygon o at time interval k,

A_o = area of polygon o,

S_o = storage coefficient of polygon o, and

Δt = finite time interval

In terms of the above parameters Cruickshank and Chavez wrote the principle of conservation of mass for their planar polygons and obtained that

$$\frac{1}{2} \sum_1^n (T_i + T_o)(h_{i,k+1} - h_{o,k+1} + h_{i,k} - h_{o,k}) \frac{B_{io}}{L_{io}} + Q_{o,k}$$

$$= \frac{A_o S_o}{\Delta t} (h_{o,k+1} - h_{o,k}) \quad (8.3\text{-}1)$$

To apply this last equation we must begin by carefully limiting the zone under consideration and locating its boundaries. Once the limits are set the zone is covered by triangles.* Finally, the polygons are obtained by drawing the bisectrices of all the angles in the triangles and their intersections will be the vertices of the polygons. The center of a polygon is the common vertex of the triangles used in obtaining the polygon.

---

* It is necessary that the triangles be such that all their internal angles be acute.

Equation (8.3-1) permits us to obtain the piezometric head in all the nodes in the zone for an instant $k + 1$, assuming or knowing the transmissivity, the storage coefficient, the piezometric head, and the discharge of every polygon at an instant k.

This model is easy to adapt to the computer and once calibrated, it can predict future behavior of an aquifer under given conditions.

Exercises

1. What is the scientific basis of the gravimetric survey? What are its advantages as far as Regional Hydrology is concerned?

2. What is a gravimeter? Consult a text on Geophysical Exploration and examine the corrections that are needed in order to adequately compare gravity values.

3. Compare the measurement of spontaneous potential, resistivity, and gamma radiation.

4. Discuss in detail the underlying principles of a seismic survey.

5. What is acid treatment? Consult the paper by Rowan mentioned in this chapter and from it evaluate the process of acidizing a well.

6. What is the purpose of casing a well?

7. Compare the simple hydrologic balance with the Cruickshank-Chávez Model.

8. What is the advantage of undertaking a thorough regional analysis prior to exploiting subsurface water resources?

CHAPTER 9

Regional Surface Hydrology

In the previous chapter we discussed how a regional groundwater study is carried out. Here, the same will be done for a regional surface water analysis. The first step in a regional surface hydrology analysis is the preliminary evaluation of a basin's surface waters and an estimation of community needs both present and future. Once this is done, the hydrologist must determine what surface water is to be used and whether or not impoundage is required.

To adequately distribute this water to individual users, adequate transmission conduits must be installed. Thought has to be given, not only to distributing water to community users, but also to satisfy the needs of irrigation and supply water to industry. At the same time, adequate fire and emergency reserves must be kept.

Another phase of regional surface hydrology is the careful, continuous evaluation of a basin. In Chapter 2 several techniques to evaluate and interpret precipitation, evapotranspiration, and runoff data were discussed. Another important technique not considered in Chapter 2 is the hydrograph. This technique was left to be discussed here and will be treated in Section 9.3. Also some applications of the hydrograph will be mentioned.

A new technique that is becoming popular in regional surface hydrology is systems analysis. For example[1] the Federal Water Pollution Control Administration carried out such a model to investigate the water quality responses in the Potomac River Basin. Their model included consideration of:

(a) low flow augmentation,
(b) wastewater diversions,
(c) water supply withdrawals, and
(d) changes in the extent of wastewater treatment.

---

[1] Hetling, Leo J. (1970) *Potomac Estuary Mathematical Model*, Water and Sewage Works, vol. 117, n. 2, pp. 47–51.

Such models are capable of handling a large number of variables and are becoming more and more useful to the regional hydrologist.

9.1 Natural and Man-made Surface Water Bodies

In Chapter 2 a discussion concerning rivers, lakes, and dams was given from a hydrologic standpoint. Now, these surface water bodies will be analyzed from a water supply point of view.

There are basically three ways of utilizing surface waters. These are:

1. Continuous Intake. Water is continuously drawn from a river or lake by way of an intake crib or tower and flows through conduits to the city's waterworks where it is treated and purified.
2. Selective Intake. Only flood waters or unpolluted waters are withdrawn from the source.
3. Water Storage or Impoundage. Upland unpolluted streams are tapped near their source and water is piped from such reservoirs to the community as demand so requires it.

The impoundage can be obtained as has been pointed out in Chapter 2 by the construction of dams and dikes. The construction of these structures is a complex engineering problem that requires numerous calculations and extensive planning. Careful consideration must be given to the selection of the point to be tapped, the size of the reservoir, the characteristics of the construction materials to be used in building the dam and the overall cost. Also, a study should be made so as to lessen the interference between water rights of individuals and the reservoir.

A dam on an impounded stream usually has several intake mechanisms to handle low as well as high level conditions. There are two basic types of dams, namely, earth dams and masonry dams.

Earth dams are usually made of a mixture of sand and gravel mixed with varying amounts of clay and silt. The upstream side of the dam must be carefully protected against erosion and wave action. The downstream side should be seeded and excess runoff should be safely intercepted and drained.

Masonry dams are complex engineering structures. Gravity dams are a type of masonry dam used when strong foundations are available. Arched dams are used in narrow valleys due to the arched action required. In plan view this type of dam looks like a bow arched to shoot an arrow. There are also numerous other types of reinforced concrete dams.

When designing a dam, great care must be taken to avoid water losses through underground seepage. These losses are very important since they can cause underground channels that may eventually undermine the dam.

Regardless of whether we are talking about river, lake, or impounded water, intake conduits should be placed in protected areas, away from traffic, and in an adequate place with respect to prevailing winds and sources of pollution.

Collection of water is not a one shot deal that once the intake structures and/or the impoundage reservoir is built can be left alone. On the contrary, much work is required in the maintenance of dams and in the control of the quality of the water that is sent to the pumping and purification works and eventually distributed to individual users.

Also, careful continuous planning is required in foreseeing times of drought and in preparing for them. In addition, expansion of existing facilities must be determined well in advance of the required time in order to avoid future water shortages.

9.2   Distribution of Surface Waters by Means of Canals

In the previous section the need for adequate means of storage of surface waters was discussed. It is just as equally important to have adequate supply conduits to transport water from the source to the user. These supply conduits are a most important part of a community's water system and must be designed in such a way as to guarantee adequate and continuous water supply to all users in the area.

The water conduits can be of two basic types: (1) Open Flow Channels, in which water is moved by gravity and the top water surface is a free surface at atmospheric pressure, and (2) Under-pressure Conduits. The open flow channels are most commonly used for land irrigation. A distribution system for irrigation purposes usually consists of a main channel furnishing water to individual lateral channels which may further subdivide into smaller ones. The main channel dominates the entire irrigable area and supplies water to all the smaller individual channels.

There are many types of conduits which can be used in the transmission of water. Some of these are dug through the ground while others are above ground. There are grade tunnels, pressure tunnels, and force mains. On occasion, depressed pipes are also used.

In designing a distribution system great care must be taken to adequately insure community water supply to all users and also to guarantee adequate fire and emergency reserves. This requires a complex plan layout which must be carefully carried by a hydrologist working closely together with a water supply specialist. Care must be taken to construct conduits of the adequate shape, carrying capacity and strength necessary for the system in question.

In designing distribution means for land irrigation one must try to estimate the following factors:

1. amount of water needed for the crop being planted,
2. water losses and other wastes,
3. irrigation methods,
4. supply sources,
5. monthly precipitation, and
6. monthly water demand.

The different distribution channels in an irrigation set-up can be located in various patterns. The easiest and usually cheapest way is to lay the channels following the topography of the land. Channels can then be located along the water divide and thus are able to dominate both sides of the divide. Also thalwegs can be used to house drains.

An important part of an efficient community water distribution system is the installation of water meters to measure individual usage and thus minimize waste. For most home users displacement type meters are adequate while current or velocity meters are more common for heavy flow rates or high pressure requirements.

The distribution and utilization of water in a community is mostly the concern of a water supply specialist. A full treatment of this subject can be found in books on waste-water disposal and water supply engineering.[2]

9.3 Evaluation of Runoff and Surface Water Optimization

In the previous sections the collection and distribution of surface waters were briefly discussed. The two remaining aspects to be treated in relation with surface hydrology are the evaluation of the physical characteristics of surface water bodies and the chemical purification of water. This last item will be considered in chapter 10 together with the subject of water pollution.

To evaluate the surface waters of a basin, one must begin by carefully limiting the basin and constructing an accurate map. Then, the following parameters must be calculated:

1. average basinwide precipitation,
2. average evapotranspiration,
3. average infiltration into the ground, and
4. average runoff.

The average precipitation in the basin can easily be determined by any of the three methods discussed in Section 2.3. Data about the frequency of precipitation is also very useful in surface hydrology evaluations.

---

[2] Fair, G. M. and Geyer, J. C. (1958) *Elements of Water Supply and Waste-Water Disposal*, John Wiley & Sons, Inc., N.Y.

The evapotranspiration represents a total basinwide water loss to the atmosphere. An average value can be obtained from data of evaporometers scattered throughout the basin (see Section 2.5).

As far as surface water bodies are concerned infiltration can be regarded as a loss of water which leaves the land's surface to replenish underground aquifers. The amount of infiltration depends on many geologic factors such as, porosity, permeability, and rock lithology. Data about these factors is very useful in a regional evaluation of surface waters.

Runoff is the portion of the water that is not lost through infiltration, nor evaporates, but goes to replenish rivers and lakes. The evaluation of the amount of runoff water is an important parameter in a research program to optimize the utilization of surface waters.

In order to identify the general characteristics of runoff in a basin, and in order to establish a basis for the analysis of its components, a graphical representation of runoff is necessary. This graphical representation is accomplished by plotting discharge of the river to which the water runs off as a function of time. Such a graph is termed a hydrograph.

The runoff value used in constructing a hydrograph must be the sum total of both the immediate and delayed runoff as both of these contribute to the discharge of a river. The importance of immediate runoff is greater the lesser the permeability of the terrain and the milder the climate.

The importance of constructing hydrographs can be appreciated by noting its applications. For example, in connection with dams, hydrographs are needed to determine:

(a) the active storage capacity of the dam,
(b) the capacity of intake conduits,
(c) the installed capacity of hydroelectric plants,
(d) the capacity of diversion channels, and
(e) the spillway capacity.

The shape of the hydrograph is determined by the characteristics of surface runoff. Analyses have been done to relate hydrograph shape with types of storm.

In runoff calculations one must include the precipitation directly over the river. However, this factor is usually small, unless the river is very big.

All the physical parameters considered above can be put together into a hydrologic inventory of the form

$$\text{Total Surface Water} = \text{Precipitation} + \text{Basin Inflow} - \text{Evap.} - \text{Infiltration} - \text{Basin Outflow}$$

Also, more complex models capable of describing certain phenomena in the basin, can be constructed in the same fashion as the Cruickshank-Chavez model discussed in the previous chapter. In brief, these models must begin by limiting the basin and dividing it into a finite number of discrete segments sufficiently small so that one can validly assume that a parameter X is constant for a given segment. For each parameter X_i one must write all the equations necessary to describe the parameter. One will need n equations where n is the number of physical quantities that affect X_i. Then, a mass balance over each of the segments in the basin is carried out much in the same fashion as in the Cruickshank-Chavez model. Usually, the process consists in solving a system of differential equations by numerical methods using a digital computer.

As for the case of regional subsurface hydrology studies, these mathematical dynamic models are most important since they allow us to forecast future conditions and to predict basin behavior under an arbitrarily given set of circumstances.

Hydrologic inventories, on the other hand, are only useful for preliminary evaluations and are only to be regarded as a simple-minded approximation to the problem.

CHAPTER 10

Regional Geochemistry of Water

10.1 Sampling Techniques

In previous chapters, both surface and subsurface waters have been discussed from a physical standpoint. The chemical aspects of hydrology will now be studied. It is important to realize that as a result of the contact of water with the geologic environment, sometimes the water acquires in solution varying amounts of both cations and anions. The amount of cations and anions contained in a sample of water can be readily and accurately determined using standardized analytical chemistry techniques. Nonetheless, the results of the analysis will be worthless if the sample was not properly taken from its source.

To sample surface waters[1] several techniques may be employed. All of these basically consist in lowering a closed plastic bottle to the desired depth, then opening the bottle, taking the sample, and finally closing and lifting the bottle. In the case of rivers and lakes, samples can be taken from a small boat. New techniques are presently being developed to sample surface waters from helicopters and small planes.

The sampling of groundwater is similar in principle to that of surface waters. Care in all cases must be taken to take the sample at the proper depth in the well.

When water is sampled, certain on-the-spot measurements are usually done. Some of the most common ones are:

1. air temperature,
2. water temperature,
3. water pH, and
4. water conductivity.

Also it is necessary to record the date the sample is taken, the location

---

[1] Rainwater, F. H. and Thatcher, L. L. (1960) *Methods for Collection and Analysis of Water Samples*, USGS WSP-1454.

of the well sampled or the location of the point in the river or lake sampled, the depth of sampling, and the name of the individual doing the sampling. All this information is important to the scientist doing the geochemical interpretation.

Present day techniques, also permit the in-situ analysis of waters. However, except for items like dissolved gases this is not absolutely necessary and the analysis can just as well be done in the laboratory.

10.2 The Most Abundant Elements in Water

Both surface waters and groundwater contain in solution many ions in varying amounts. The concentration of each of these ions can be determined, in most cases, rather easily by analytical means. These ionic concentrations are most commonly expressed in terms of parts per million by weight. Also, concentrations are expressed at times in terms of milligrams per liter of solution.

If one desires to compare concentrations of various ions in the water, it is more convenient to use equivalents per million. To convert to equivalents per million from parts per million one must use the formula:

equivalents per million = ppm × ionic charge/atomic or molecular weight

There are other units of measurement which are occassionally used in expressing concentrations of ions in solution. These units compare to parts per million as follows:

unit	ppm
mg/liter	1.00
grains per gallon (U.S.)	17.12
grains per gallon (Imp.)	14.27
French degree	10.00
German degree	17.80

The most important chemical constituents present in water are: sodium, calcium, magnesium, potassium, silica, chloride, fluoride, nitrates, sulfates, and carbonates. A discussion of each follows.

SODIUM

Of all metallic ions, sodium is the most abundant in sea water. Concentrations of 8,000–14,000 ppm are common in ocean water and the sodium variations are due to salinity changes. On the other hand, sodium does not appear in large amounts in groundwater except in connate waters (water of deposition) or when salt water from the ocean intrudes into a given aquifer. Nonetheless, some sodium is always present in most groundwaters. Sources of sodium in groundwater are evaporate rocks such as halite (rock salt), or igneous rocks. Once sodium is put in solution it tends to remain as such and will not precipitate under usual aquifer conditions.

Sodium is easily analyzed by spectrophotometric means.

CALCIUM

Calcium is a very important constituent of most waters. It originates from dissolution of rocks such as limestone, calcite, aragonite, anhydrite, anorthite, wollastonite, or chalk in which calcium is an essential component. When water percolates through such rocks it dissolves calcium in the form of Ca^{++}. The presence of calcium in large amounts requires the presence of large amounts of dissolved CO_2.

Calcium can be analyzed spectrophotometrically, however, sodium interferes. It is common now to analyze calcium by the EDTA method. This method is discussed in reference 1 of this chapter.

MAGNESIUM

Magnesium is abundant in waters from dolomitic aquifers. Similar to calcium, magnesium is present in solution in ionic form, however, it has a greater tendency than calcium to remain in solution.

For most waters the ratio of calcium to magnesium in equivalents per million varies between 5:1 and 1:5. High values (5:1) with respect to calcium suggest that the water has been in contact with calcium carbonate. Values of the order of 1:1 or 1:2 suggest dissolution of olivines or dolomites. High values (1:5) with respect to magnesium represent cases of salt water intrusion.

POTASSIUM

Potassium is an important metallic ion in water. Waters associated with igneous silicic rocks contain sodium and potassium in equivalent amounts. Also, these waters possess relatively low concentrations of total dissolved solids. In most surface and subsurface waters the potassium concentration is low, however, thermal waters usually possess higher concentrations of potassium.

SILICA

Groundwater usually contains from 10 to 40 parts per million of silica. Rarely are concentrations of over 200 ppm SiO_2 observed. Silica in groundwater usually comes from the dissolution of silicates. The amount of silica in solution will depend on the pH of the water, its temperature, and the velocity of flow. These are the basic factors that control the solubility of silica.

CHLORIDE

Chloride is the most abundant anion in sea water. On occasions it occurs in amounts exceeding 20,000 ppm. In rain water Cl^- varies from 1 to 10 ppm, and in groundwater it usually does not exceed 200 ppm. For drinking purposes water should not have over 225 parts per million of chloride. Waters having in excess of 500 parts per million of Cl^- have a rather unpleasant taste.

FLUORIDE

Fluoride occurs in very small quantities in groundwater. It is usually derived from fluorite and other fluoride bearing minerals. Fumarolic and volcanic gases also contain some fluorides. An excess of fluorides in water used for drinking is dangerous for children because it tends to cause mottled teeth.

NITRATES

Nitrates appear in groundwater as the result of either sewage contamination or passage through alfalfa-planted soil. Waters with high concentrations of nitrates usually contain bacteria. Concentrations in excess of 50 ppm nitrate ion are not suitable for drinking purposes as they may cause cyanosis.

SULFATES

The sulfate ion in groundwater is usually derived from contact with gypsum or anhydrite. When water contains large amounts of sulfates it has a laxative effect, especially on people not used to drinking it.

CARBONATES

Carbonates and bicarbonates are present in water in varying amounts. Most common ones are those of calcium and magnesium. Both calcium and magnesium carbonate cause temporary water hardness. This hardness can be removed by boiling the water and precipitating $MgCO_3$ and $CaCO_3$.

10.3 Water Pollution

Due to the rapid industrial and population growth that some metropolitan areas have experienced in the past few years, many rivers, streams, lakes, and aquifers have been loaded to such high concentrations of certain cations and anions that their waters are no longer suitable for human consumption, nor can they (in the case of surface waters) support fish and wildlife.

There are numerous types of pollution and all cannot be adequately treated in the short space here devoted to pollution. There is natural pollution, as the result of percolation of groundwater through natural sedimentary strata containing cations and anions which can sorb to the water. This is the least worrisome type of pollution. Sea-water intrusion is a troublesome phenomenon that makes water in some coastal aquifers unfit to drink without being subjected to treatment. Mercury, especially in the methyl form, has contaminated many surface water bodies. Radioactive wastes and pesticides are also very dangerous. Finally, phosphates, which have been in the past extensively used in detergents because of their surfactant action, have the

Table 10.1 Some industrial contaminants.

Contaminant	Suggested max. level in drinking water (mg/liter)
ABS	0.5
Ammonium	0.5
Arsenic	0.05
Barium	1.0
Cadmium	0.01
Chromium	0.05
Copper	1.0
Cyanide	0.05
Fluoride	1.5
Lead	0.05
Mercury	0.005
Nitrates	50
Phenol	0.001
Selenium	0.01
Silver	0.05
Zinc	5.0

ability to accelerate plant growth and thus cause changes in the oxygen balance of a river or lake. This in turn may cause death of fish and animal life in the stream or lake. Lake Erie is a case in point.

Modern industry uses numerous substances which are eventually dumped into streams and aquifers as wastes, and which are dangerous to human beings and animals. Some such substances are listed in Table 10.1. Suggested maximum levels of these substances in drinking water are also listed.

Increasing water pollution and consequent water shortages show that both municipalities and industries should attempt to optimize their utilization of water and minimize pollution. Thus, water must be re-used as much as possible. It has been rightly pointed out by the Water Pollution Control Federation that wastewater more and more represents an increasing portion of the resources of a nation and should be advantageously used.

Wastewater reclamation can be achieved in various ways depending on the utilization to be given to the recleaned water. In any reclamation project, the following five objectives should be kept in mind:

(i) optimization of water usage,
(ii) minimization of waste,
(iii) prevention of nuisance conditions,
(iv) protection of health, and
(v) conservation of natural resources.

The cleanest water should be kept for human consumption, while recleaned water is supplied to industry and used for irrigation. A prime example of the beneficial aspects of wastewater utilization is the irrigation with wastewater that is done in some agricultural areas utilizing urban sewage water from nearby cities. Of course, care must be taken not to contaminate or damage the crops. Blosser and Owens (1964)[2] discuss the land disposal of pulp mill effluents and the subsequent effect on agriculture.

Studies have been made to see how natural sediments may be used in part to clean up the pollution that may result from usage of non-biodegradable detergents[3] or from sulfonate flotation.[4]

The use of naturally available sedimentary beds is cheap and may permit utilization of certain amount of wastewater to be more economically favorable. Of course, the use of this technique would have to be coupled with other processes which may be needed in cleaning a given wastewater. Some of these other techniques will now be discussed.

Adsorbents, especially activated charcoal, have been used and found effective. Middleton[5] states that preliminary cost projections for reduction of 50–70 ppm of organic materials to 10–15 ppm are from 5–10¢ per 1000 gallons of wastewater for plants handling more than 10 mgd. This figure includes both amortization and operating costs.

---

[2] Blosser, R. O. and Owens, E. L. (1964) *Irrigation and Land Disposal of Pulp Mill Effluents*, presented at 11th Annual Ontario Industrial Wastes Conference, reprinted Water and Sewage Works, 1970, Water Re-Use the Objective of Advance Waste Treatment Research, pp. 41–49.

[3] Deju, R. A. (1970) *A Method for Wastewater Clean Up Using Natural Sediments*, Publication 1038, Institute of Geophysics, University of Mexico, pp. 1–22.

[4] Davis, F. T. (1970) *The Measurement and Control of Water Pollutants From Flotation Plants*, paper presented at the Pacific Southwest Mineral Industry Conference, San Francisco, California.

[5] Middleton, F. M. (1970) *Advanced Treatment of Wastewaters for Re-Use*, in Water Re-Use . . . The Objective of Advanced Waste Treatment Research, Scranton Gillete Pub. Co., pp. 18–27.

Ion exchange resins are also available for water reclamation, and they come in various forms. These can be used for final water quality polishing. Many laboratories now use ion exchange columns and are able to obtain almost zero conductivity water. Ion exchange resins have also been used for water softening.

Distillation is another technique for water cleanup. However, distillation does not ordinarily produce a drinkable product since volatile contaminants and low boiling compounds appear in the distillate. Distillation of seawater is nowadays a well known process. Also much thought has been given during the past few years to the construction of joint plants for wastewater distillation and power production.

Solvent extraction is one of the most interesting ways to clean up water. The technique can be outlined as follows:

1. contaminated brackish water $+$ amine solvent $\rightarrow$ brine $+$ solvent-H_2O

2. solvent-H_2O $\xrightarrow[\Delta T]{}$ solvent $+$ clean water

3. The clean water is finally air stripped.

As indicated above, the process is basically an organic separation followed by air stripping of the clean water effluent.

This process is very effective for the removal of organic salts. However, it is rather expensive.

There are numerous other methods for treating wastewater. Some of the most important ones are filtration through sand and gravel, foam separation, chlorination, reverse osmosis, freezing, aeration, oxidation of organic contaminants to carbon dioxide, polyelectrolyte addition, diatomaceous earth filtration, and alum and lime treatment.

10.4   Properties of Water

In addition to the determination of cations and anions present in water, it is useful to determine certain additional properties. Most important of these are hardness, pH, conductivity, and total dissolved solids.

HARDNESS

The hardness of water is due to the presence of carbonates, sulfates, chlorides, and nitrates of calcium and magnesium. The hardness of a particular water will influence its sudsing capacity. In excessively hard water soap does not produce suds.

There are two basic types of hardness:

a. temporary hardness, which is due to the presence of carbonates and bicarbonates, and can be removed by boiling the water and precipitation of $CaCO_3$ and $MgCO_3$, and

b. permanent hardness, which is due to the presence of $CaCl_2$, $MgCl_2$, $CaSO_4$, $MgSO_4$, $CaNO_3$, and $MgNO_3$. These can not be eliminated by boiling the water.

The hardness of water can be calculated by the equation

$$\text{ppm hardness as } CaCO_3 = (\textstyle\sum \text{epm Ca} + \text{Mg} + \text{Ba}) \times 50.05$$

It can also be determined in the laboratory using the EDTA method (reference 1 of this chapter).

Waters with hardness of less than 50 ppm are considered soft. Those waters where the hardness exceeds 150 or 200 ppm must be softened before using them.

pH

The pH of an aqueous solution is defined by the expression

$$pH = -\log [H^+]$$

where $[H^+]$ is the hydrogen ion concentration in the solution. Water with a pH of 7.0 is neutral. Waters with pH values below 7.0 are said to be acid, and those with pH above 7.0 are said to be basic. The majority of natural waters have pH values between 5.5 and 7.5. The pH of the water is greatly influenced by the rocks it has been in contact with.

Groundwater contains, in most instances, carbon dioxide and other dissolved gases. The solubility of these depends on both pressure and temperature. When a sample of groundwater is taken, the pressure and temperature change. Thus, the solubility of the dissolved gases changes and the pH varies. This is the reason why it is so important to measure the pH rapidly in the field before further variations take place. The pH as measured in the laboratory several days later may considerably differ from that measured in the field.

CONDUCTIVITY

The property of a solution of conducting electricity is its conductivity. Distilled water has nearly zero conductivity and the more ions are present in the water, the greater the conductivity. Thus, conductivity can be used as an index of water quality and total amount of dissolved minerals.

The unit of conductivity is the mho, which is the inverse of the ohm. Since mhos are relatively large, it is more practical to express the conductivity of water in terms of micromhos, that is 10^{-6} mhos. Conductivity can easily be measured in the field or in the laboratory using a conductivity bridge. Numerous models of these bridges are commercially available.

TOTAL DISSOLVED SOLIDS

The amount of minerals dissolved in a sample of water is an important parameter and must be determined. This determination is usually done by weighing the dry residue of the evaporated water sample.

The total dissolved solids can also be calculated by summing the concentrations of the different ions found in the sample. This method is not as accurate as the preceding one.

Waters possessing over 1000 ppm total dissolved solids must be treated before using them. In most instances, waters with over 500 ppm total dissolved solids are not regarded as potable.

10.5   Techniques in Basinwide Geochemical Interpretation

The key part of a geochemical survey is the correlation and interpretation of all the data obtained. To adequately carry this job the geohydrologist needs to have for every data point the following determinations:

pH	Na^+	total hardness	conductivity
SO_4^{--}	K^+	SiO_2	NO_2^-
Cl^-	Li^+	B	NO_3^-
CO_3^{--}	Ca^{++}	total dissolved solids	PO_4^{-4}
HCO_3^-	Mg^{++}	total acidity	temperature of H_2O

On the basis of the chemical determinations mentioned above, the geohydrologist can proceed to:

(a) calculate the salinities and alkalinities of the waters analyzed,
(b) correlate and differentiate waters on the basis of their geochemical index, and
(c) classify the waters according to the system of Chase Palmer.

These three steps help in elucidating the origin of groundwater in a basin, determine the chemical nature of aquifer rocks, and determine approximately (if enough data points are available) the nature of the underground flow regime. All this information is extremely valuable in a regional groundwater survey.

In order to carry points (a), (b), and (c) above one must begin by performing the following calculations (in epm):

$$
\begin{aligned}
\text{Sum of Free Acids} &= \sum SO_4^{--} + Cl^- + NO_3^- + NO_2^- = FA \\
\text{Sum of Weak Acids} &= \sum CO_3^{--} + HCO_3^- + PO_4^{-4} \quad = WA \\
\text{Sum of Alkalies} &= \sum Na^+ + K^+ + Li^+ + NH_4^+ \quad = A \\
\text{Sum of Earths} &= \sum Ca^{++} + Mg^{++} \quad = E
\end{aligned}
$$

On the basis of the four sums calculated above, one can classify the waters according to Palmer's system. This classification is as follows:

$$
\begin{aligned}
\text{Class I} &\quad FA < A \\
\text{Class II} &\quad FA = A \\
\text{Class III} &\quad FA > A \text{ and } FA < A + E \\
\text{Class IV} &\quad FA = A + E \\
\text{Class V} &\quad FA > A + E
\end{aligned}
$$

Once the class has been determined one can proceed to calculate the salinities and alkalinities. These quantities are defined as follows:

$S1$: The primary salinity in a water is the amount of alkalies that are balanced by free acids, that is, the amount of salts resulting from the union of alkalies with free acids.

$S2$: The secondary salinity is the amount of alkaline earths balanced by free acids.

$S3$: The tertiary salinity is the excess of free acids over alkalies and alkaline earths. Waters where $S3 > 0$ have free acidity.

$A1$: The primary alkalinity in a water is the amount of alkalies that are balanced by weak acids.

$A2$: The secondary alkalinity is the amount of alkaline earths that are balanced by weak acids.

$A3$: The tertiary alkalinity is that due to the presence of heavy metals. Most waters have $A3 = 0$. If $A3 > 0$ some nearby mineralization must be present. The most common heavy metals causing tertiary alkalinity are copper, lead, zinc, iron, and manganese. The letter M will denote the sum of heavy metals.

The values of the salinities and alkalinities vary depending on the class (Palmer) of the water. They can be calculated according to the following rules:

Class I	Class II
$S1 = 2FA$	$S1 = 2FA$
$A1 = 2(A - FA)$	$A2 = 2E$
$A2 = 2E$	$A3 = 2M$
$A3 = 2M$	

Class III	Class IV
$S1 = 2A$	$S1 = 2A$
$S2 = 2(FA - A)$	$S2 = 2E$
$A2 = 2(A + E - FA)$	$A3 = 2M$
$A3 = 2M$	

Class V
$S1 = 2A$
$S2 = 2E$
$S3 = 2(FA - A + E)$
$A3 = 2(M - S3/2)$

The salinities and alkalinities can be used to explain certain characteristics of the water. If the primary alkalinity is the predominant value, then the water must have percolated through sedimentary alkali beds. If the secondary alkalinity predominates, the water percolates through sedimentary strata rich in alkaline earths. Usually waters where $A2$ is predominant possess temporary hardness. When $A3$ predominates, it implies that the water passes through mineralized strata and possesses free acidity.

If the primary salinity is predominant, it implies that the water passes through volcanic terrain. If $S2$ predominates the water is usually one that has passed through sedimentary strata, volcanic in origin. Generally such waters have permanent hardness. Finally, if $S3$ predominates, the water must have percolated through mineralized terrain. Such waters usually possess free acidity.

In addition to the Palmer classification and to the calculation of salinities and alkalinities, some geohydrologists like to classify waters according to the Geochemical Index. This index is defined by the equation:

$$G.I. = FA + E - 50\%$$

There are also several graphical techniques which are helpful in the geochemical interpretation of a basinwide groundwater system. The most important ones are:

1. basin concentration maps,
2. semilogarithmic plots, and
3. triangular plots.

If there are enough data points in the basin, then, contours of ionic concentration of certain ions can be plotted. Such maps will indicate any zone differentiation within the basin. Lines of equal concentration of boron and Cl are commonly used. Boron is found in tourmaline, pegmatites, granites and in smaller quantities in volcanic rocks such as gabbro. It can be liberated from volcanic gases as H_3BO_3 or BF_3.

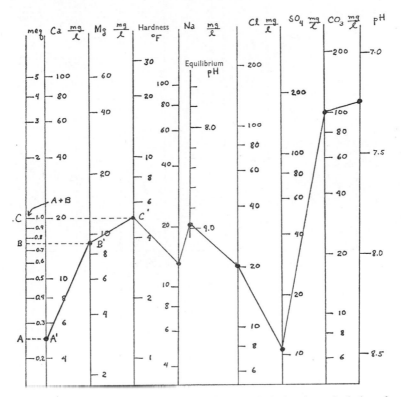

Figure 10.1 Semilogarithmic plot of a water analysis showing calculation of hardness.

Water from fumaroles and areas that recently underwent volcanic activity have a high boron content. Also boron is indicative of faults at depth.

Chlorine is present in most waters. Those waters associated with sedimentary rocks possess between 10 and 40 ppm Cl$^-$. Low values of Cl$^-$ associated with low sulfate ion concentration indicate an igneous origin. High chloride concentrations are usually associated with salt-water intrusion or marine formations.

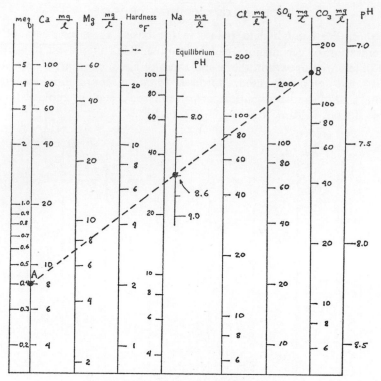

Figure 10.2 Calculation of equilibrium pH.

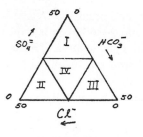

(A) Cations

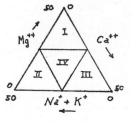

(B) Anions

Figure 10.3 Triangular plots.

188

Semilogarithmic diagrams are also very useful. These are simply nomographs which automatically convert ppm of various ions into milliequivalents. This type of graph contains scales for calcium, magnesium, hardness, sodium, chloride, sulfate, carbonate, pH, and equilibrium pH (see Figure 10.1). The hardness can be directly calculated in such a diagram as follows. First transfer A' to A and B' to B. Then add $A + B = C$ and finally transfer C to C'. This last value will be the hardness in French degrees. The entire computational process is shown on Figure 10.1.

The equilibrium pH can also be calculated easily from a semilogarithmic plot. The equilibrium pH is obtained by connecting points A and B in Figure 10.2. The equilibrium pH is that obtained when $CaCO_3$ equilibrium is achieved. In Figure 10.2 the equilibrium pH is 8.6. If the equilibrium pH is less than the actual pH of the sample, deposition of $CaCO_3$ is possible.

The triangular or trilinear plots are representations of the percent of certain ions in solution. Figure 10.3A shows the diagram for the cations. Region I corresponds to a magnesium-rich water, region II to a soda-rich water, region III to a calcium-rich water and region IV to mixed water. Similar comments can be made in relation to the diagram for the anions shown in Figure 10.3B. Both of these diagrams can be added and plotted in a composite rhombic diagram.

In addition to all the formulas and diagrams mentioned above, a most important part of the interpretation is the experience and knowhow of the regional hydrologist.

Conversion Factors

To convert from:	Into:	Multiply by:
acres	meters^2	4047.00
acre-feet	gallons (US)	3.26×10^5
atmospheres	cm of merc.	76.0
oil barrels	oil gallons	42.0
centimeters	feet	3.28×10^{-2}
centimeters	inches	0.39
cubic centimeters	gallons (US)	2.64×10^{-4}
cubic feet/min	gal/sec	0.125
cubic feet/sec	gal/min	449.0
cubic feet/sec	mgd	0.65
cubic meters	gallons (US)	264
darcy	centimeters^2	1×10^{-8}
feet	centimeters	30.5
feet of water	atmospheres	0.03
gallons (US)	gallons (IMP)	0.83
gallons per day/feet^2	cm/sec	4.65×10^{-5}
kilometers	miles	0.62
kilometers/hour	feet/sec	0.91
liters	gallons (US)	0.26
liters/minute	feet^3/sec	5.9×10^{-4}
meters	miles (st)	6.21×10^{-4}
square meters	acres	2.47×10^{-4}

FORMULA CONVERSIONS

To find the formation constants using Theis' method formulas (7.2-1) and (7.2-2) were used. It is customary in the United States to also write these formulas as

$$T = \frac{114.6QW(u)}{s}$$

and

$$S = \frac{0.535uTt}{r^2}$$

where s is expressed in feet, Q in gallons per minute, T in gallons per day/ square feet, r in feet and t in days since the beginning of pumping.

In Europe and countries other than the United States and Canada formula (7.6-1) is expressed in the form

$$t_{nc} = 10.25 \times 10^3 \, Rb/K$$

where t_{nc} is in hours, b is in meters, K is in cm/sec and R is dimensionless.

PPM-EPM Conversion Table

Conversion factors: Parts per million to equivalents per million

Ion	Multiply by
Aluminum +3	0.11119
Bicarbonate	0.01639
Calcium +2	0.04990
Carbonate	0.03333
Chloride −1	0.02820
Fluoride −1	0.05263
Hydrogen	0.99206
Hydroxide	0.05880
Iron +2	0.03581
Iron +3	0.05372
Lithium +1	0.14409
Magnesium +2	0.08224
Manganese +2	0.03640
Nitrate	0.01613
Potassium +1	0.02558
Sodium +1	0.04350
Sulfate	0.02082

APPENDIX 3

Water Laws in the United
States of America*

Water is an important asset to both individuals and municipalities and as such it has value. Its value depends on many factors and greatly depends on the abundance of water in the area. Water rights guarantee individuals the right to use water for a particular use and in a specified quantity. Water rights are classed as property rights and guarantee the right to use water but do not give ownership of the water body. In many instances conflict occurs between various users and the law has to enter into the picture. Water laws are different from one state to another, nonetheless some useful generalizations can be made.

Water rights are granted on the basis of various criteria by the state governments. Most water laws do not insure an optimization of water usage. Garrity and Nitzschke in their Water Law Atlas[1] state that:

" If a water right could be used anywhere within a watershed or basin and not restricted to a specific parcel of land, the open market will insure its most economic use."

However, water laws have not yet moved in this direction.

When one considers the legal aspects of water a clear distinction must be made between surface waters, underground streams, and percolating groundwater. These are defined as follows:

(i) Surface waters are those that run on definite channels along the earth's surface and that possess a surface in contact with the atmosphere.

---

* Material from the *Water Law Atlas*, Circular 95, State Bureau of Mines and Mineral Resources, New Mexico Institute of Mining and Technology has been used in compiling this appendix. This material has been used with permission from New Mexico Institute of Mining and Technology, Socorro, New Mexico, 87801.

[1] Garrity Jr. T. A. and Nitzschke Jr. E. T. (1968) *Water Law Atlas*, Circular 95, State Bureau of Mines and Mineral Resources, New Mexico Institute of Mining and Technology, Socorro, New Mexico, 87801.

(ii) Underground streams are well-defined channels of water that underlie certain streams and contribute to surface flow.

(iii) Percolating groundwater is that portion of the water that precipitates on the earth and filters through the ground, moving slowly through the pores of the subsoil.

The laws governing surface waters and underground streams are the same and will be considered first. Laws regulating percolating groundwater will be studied separately.

There are four principles that apply as far as surface waters and underground streams are concerned. Some states follow only one while in others more than one doctrine is applied. These are:

1. *Riparian common law doctrine.* According to this principle the right to use water is inseparably attached to the ownership of the soil. The principle dates back from the beginning of common law. It is mostly applied in the eastern part of the United States. This principle does not permit the consumptive use of water.

2. *Riparian reasonable use doctrine.* This is a modification of the preceding principle guaranteeing that a certain amount of water can be used for consumptive purposes.

3. *Appropriation principle.* This is the most widely used doctrine in the western United States and it is based on the rule that the first one to appropriate and use the water has a right superior to anyone coming later.

4. *Principle of need.* The most needy, according to this rule, has the right to use water. This law is mostly applied to Indian water rights.

Table 1 lists the states of the Union and indicates which of the above doctrines applies in each of them. Data contained in this table was compiled from reference 1 of this appendix.

As far as percolating groundwater laws are concerned four basic principles need be defined. These are:

1. *Riparian right of absolute ownership.* In states where this principle applies it entitles the bearer and owner of the land to withdraw from his wells any amount of groundwater he desires for whatever purpose he sees fit.

2. *Riparian right of reasonable use.* This is a limitation on the preceding principle entitling the owner of the land to withdrawing water in amounts so as not to damage the rights of other users as far as quantity and quality.

Table 1 Laws applicable to surface waters and underground streams.

State	Common law riparian	Reasonable use riparian	Appropriation doctrine
Alabama		×	
Alaska			×
Arizona			×
Arkansas	×		
California		×	×
Colorado			×
Connecticut	×		
Delaware	×		
Florida		×	×
Georgia	×		
Hawaii	×		×
Idaho			×
Illinois		×	
Indiana	×		
Iowa		×	
Kansas		×	×
Kentucky		×	
Louisiana	×		
Maine	×		
Maryland		×	
Massachusetts	×		
Michigan		×	
Minnesota	×		
Mississippi		×	×
Missouri	×		
Montana			×
Nebraska		×	×
Nevada			×
New Hampshire		×	
New Jersey	×		
New Mex.			×
New York		×	
North Car.		×	
North Dak.		×	×
Ohio		×	
Oklahoma		×	×
Oregon		×	×
Pennsylvania	×		
Rhode Is.	×		
South Car.	×		
South Dak.	×		×
Tennessee		×	
Texas		×	×
Utah			×
Vermont	×		
Virginia		×	
Washington			×
West Virginia	×		
Wisconsin		×	
Wyoming			×

Table 2 Groundwater laws.

State	Common law riparian	Reasonable use riparian	Appropriation doctrine
Alabama	×		
Alaska			×
Arizona		×	
Arkansas	×		
California		×	
Colorado			×
Connecticut	×		
Delaware	×		
Florida		×	
Georgia	×		
Hawaii	×		
Idaho			×
Illinois		×	
Indiana	×		
Iowa		×	
Kansas			×
Kentucky		×	
Louisiana	×		
Maine	×		
Maryland		×	
Massachusetts	×		
Michigan		×	
Minnesota	×		
Mississippi	×		
Missouri	×		
Montana			×
Nebraska		×	
Nevada			×
New Hampshire		×	
New Jersey	×		
New Mexico			×
New York		×	
North Carolina		×	
North Dakota			×
Ohio	×		
Oklahoma			×
Oregon			×
Pennsylvania	×		
Rhode Island	×		
South Car.	×		
South Dak.			×
Tennessee		×	
Texas	×		
Utah			×
Vermont	×		
Virginia		×	
Washington			×
West Va.		×	
Wisconsin	×		
Wyoming			×

3. *Principle of correlative rights.* This principle is in force in the State of California and it permits reasonable use of groundwater with the proviso that in times of water shortage the rights of all users must be correlative.

4. *Appropriation principle.* According to this law the first to appropriate has the right to use the water.

Table 2 contains a summary of groundwater laws in the United States. This table was also compiled from data in reference 1.

Certain states have orders of preference depending on the use to be given to the water. While these vary from state to state it is common to give preferred status to domestic consumption, and then municipal uses, irrigation, industrial uses, recreation, and finally power production.

The water situation is more critical in the western states than in the east. The short supply of water in the west has so far been a cause hindering its growth. Most western states have established weather modification legislation and carry and support "rainmaking research."

With the present emphasis on the environment and the well publicized pollution that exists in many rivers and lakes, laws have been established by all states and the federal government to enforce pollution control. These laws vary considerably from state to state.

Author Index

Subject Index